Progress in Nonlinear Differential Equations and Their Applications
Volume 5

Composite Media and Homogenization Theory

An International Centre for Theoretical Physics Workshop Trieste, Italy, January 1990

Edited by
Gianni Dal Maso
Gian Fausto Dell'Antonio

With 34 Illustrations

1991

Birkhäuser
Boston · Basel · Berlin

Gianni Dal Maso
Scuola Internazionale Superiore di Studi Avanzati
Trieste, Italy

Gian Fausto Dell'Antonio
Dipartimento di Matematica
Università "La Sapienza"
Roma, Italy

Library of Congress Cataloging-in-Publication Data
Composite media and homogenization theory : an International Centre for Theoretical Physics workshop, Trieste, Italy, January 1990 / Gianni Dal Maso, Gian Fausto Dell'Antonio, editors.
p. cm. — (Progress in non-linear differential equations and their applications ; 5)
"Proceedings of the Workshop on Composite Media and Homogenization Theory held in Trieste, Italy, from January 15 to 26, 1990. The workshop was organized by the International Centre for Theoretical Physics...; part of the activity was co-sponsored by the International School for Advanced Studies"—Pref.
Includes bibliographical references.

1. Continuum mechanics—Congresses. 2. Differential equations, Partial—Congresses. I. Dal Maso, Gianni. II. Dell'Antonio, G. F., 1933- . III. International Centre for Theoretical Physics. IV. International School for Advanced Studies (Trieste, Italy) V. Workshop on Composite Media and Homogenization Theory (1990 : Trieste, Italy) VI. Series: Progress in nonlinear differential equations and their applications ; v. 5.
QA808.2.C665 1991
531—dc20 90-19969

Printed on acid-free paper.

Softcover reprint of the hardcover 1st edition 1991

ISBN-13: 978-1-4684-6789-5 e-ISBN-13: 978-1-4684-6787-1
DOI: 10.1007/978-1-4684-6787-1

Camera-ready copy prepared by the authors.

9 8 7 6 5 4 3 2 1

Preface

This volume contains the Proceedings of the Workshop on Composite Media and Homogenization Theory held in Trieste, Italy, from January 15 to 26, 1990. The workshop was organized by the International Centre for Theoretical Physics (ICTP); part of the activity was co-sponsored by the International School for Advanced Studies (SISSA).

The workshop covered a broad range of topics in the mathematical theory of composite materials and homogenization. Among the specific areas of focus were homogenization of periodic and nonperiodic structures, porous media, asymptotic analysis for linear and nonlinear problems, optimal bounds for effective moduli, waves in composite materials, optimal design and relaxation, random media.

The workshop was actively attended by more than 100 participants from 23 countries. In the afternoon sessions 35 seminars were delivered by the participants.

This volume contains research articles corresponding to 14 of the 20 invited talks which were presented. Its content will be of interest both to mathematicians working in the field and to applied mathematicians and engineers interested in modelling the behaviour of composite and random media.

We are pleased to express here our thanks to the ICTP for having made this workshop possible, to Ms. A. Bergamo for her continuous help during the workshop, and to Ms. C. Parma for her collaboration in editing the proceedings.

Gianni Dal Maso
SISSA, Trieste

Gian Fausto Dell'Antonio
Università "La Sapienza", Roma

Contents

List of Speakers

Invited speakers

M. AVELLANEDA (Courant Institute, New York, USA): Homogenization and scaling for convection-diffusion equations.

N.S. BAKHVALOV (USSR Academy of Sciences, Moscow): Properties of averaged models of periodic media mechanics.

A. BENSOUSSAN (INRIA, Le Chesnay, France): Homogenization of a class of stochastic partial differential equations.

D.J. BERGMAN (Tel Aviv University, Israel): Weakly and strongly nonlinear (power law) electrical properties of isotropic composite media.

L. BOCCARDO (Rome University "La Sapienza", Italy): Homogenization of quasi-linear equations

G. BUTTAZZO (University of Ferrara, Italy): Relaxed formulation for a class of shape optimization problems.

A. CHERKAEV (Academy of Sciences, Leningrad, USSR): An approach to the G_m closure problem for two-component composites.

E. DE GIORGI (Scuola Normale Superiore, Pisa, Italy): New problems in Γ-convergence.

U. HORNUNG (Universität der Bundeswehr, München, F.R. Germany): Homogenization of flow and transport through porous media.

E.Ya. KHRUSLOV (Ukrainian Academy of Sciences, Kharkov, USSR): Homogenized models of composite media.

N. KIKUCHI (University of Michigan, Ann Arbor, USA): Topology and shape optimization of an elastic structure using the homogenization method.

R.V. KOHN (Courant Institute, New York, USA): Variational models of coherent phase transitions.

S.M. KOZLOV (Moscow Institute of Civil Engineering, USSR): Geometry and asymptotitcs in homogenization.

G.W. MILTON (Courant Institute, New York, USA): The field-equation recursion method.

U. MOSCO (Rome University "La Sapienza", Italy): Composite media and Dirichlet forms.

F. MURAT (Université de Paris VI, France): Correctors for the wave equation.

O.A. OLEINIK (Moscow State University, USSR): Spectral properties of operators singularly depending on parameters and applications to mathematical physics.

G.C. PAPANICOLAOU (Courant Institute, New York, USA): Wave propagation in random media.

P. SUQUET (Laboratoire de Mècanique et d'Acoustique, Marseille, France): Plasticity, yield design, and homogenization.

S. VARADHAN (Courant Institute, New York, USA): Non equilibrium problems, hydrodynamic scaling, large deviations and homogenization.

Seminar Sessions

K. ADEROGBA (University of Lagos, Nigeria): The overall physical properties of a solid reinforced by parallel plates.

A.P. AKINOLA (Obafemi Awolowo University, Ile-Ife, Nigeria): On homogenization and large deformations.

G. ALBERTI (Scuola Normale Superiore, Pisa, Italy): Lusin type theorem for gradients.

M. ARTOLA (Commissariat a l'Enérgie Atomique, Le Barp, France): Electromagnetic wave propagation in periodic media.

M. AVELLANEDA (Courant Institute, New York, USA): Compactness methods in homogenization.

S. BALDO (Scuola Normale Superiore, Pisa, Italy): A minimality condition for magnetisation domains.

G. BOUCHITTE (Université de Toulon et du Var, France): Limit analysis of the Maxwell system in case of a very thin scattering body.

V. CHIADÒ PIAT (SISSA, Trieste, Italy): Homogenization of monotone operators.

R. DE ARCANGELIS (Salerno University, Italy): The Lavrentiev phenomenon as a source of different homogenization processes.

A. DEFRANCESCHI (SISSA, Trieste, Italy): G-convergence of monotone operators.

P. DONATO (Naples University, Italy): Homogenization with small shape-varying perforations.

R. FIGARI (Naples University, Italy): Mixed boundary conditions on perforated domains and point interactions in quantum physics.

G. FRANCFORT (Laboratoire Central des Pontes et Chaussées, Paris, France): A remark on the structure of Stokes equation and application to homogenization.

J. FRANCU (Technical University of Brno, Czechoslovakia): Homogenization and correctors for nonlinear elliptic equations.

L. GIBIANSKI (Academy of Sciences, Leningrad, USSR): The estimates of effective moduli of two-component elastic composites.

K.M. GOLDEN (Princeton University, USA): Critical phenomena in random resistance network.

R.P. LIPTON (University of California, Berkeley, USA): Bounds on elastic moduli for two phase anisotropic elastic composites.

M.L. MASCARENHAS (Matematica e Aplicaçaoes Fundamentais, Lisbon, Portugal): A Galerkin approximation method in multicellular beams.

A. MIKELIC (Rudjer Boskovic Institute, Zagreb, Yugoslavia): Homogenization of two-component miscible flows through porous media.

V. NESI (Heriot-Watt University, U.K.): Remarks on the G-closure problem.

D. S. PAL (Punjab Agricultural University, Ludhiana, India): Effects of blood flow, curved boundary and environmental conditions on temperature distribution in a two dimensional structure of human skin and subcutaneous tissues.

D. PERCIVALE (SISSA, Trieste, Italy): Perfectly plastic plates surrounded by soft material.

C. PICARD (Université d'Amiens, France): On the homogenization of foliated annuli.

R.B. TAO (Fudan University, Shanghai, P.R. China): Calculation of effective physical parameters in porous and composite media with periodic structure.

N.A. TCHOU (Rome University "La Sapienza", Italy): Semilinear elliptic equations with singular potential.

J.J. TELEGA (Polish Academy of Sciences, Warsaw, Poland): On some results of homogenization: piezoelectricity, fissured solids and plates, shells.

P. VAN BEEK (Delft University of Technology, The Netherlands): A solution scheme for many-particle interactions based on the method of reflections.

R. Y. VASUDEVA (Andhra University, Waltair, India): On a model of weak adhesion in composite plate vibrations.

E. VITALI (SISSA, Trieste, Italy): Convergence of solutions of variational inequalities in higher order Sobolev spaces.

Follow-up activity, S.I.S.S.A., Trieste
January 29 - 31, 1990

M. AVELLANEDA (Courant Institute, New York, USA): Multi-scale analysis of composite media.

G. FRANCFORT (Laboratoire Central des Pontes et Chaussées, Paris, France): The G-closure for two-dimensional elasticity.

K.M. GOLDEN (Princeton University, USA): Homogenization of random electric resistors.

S.M. KOZLOV (Moscow Institute of Civil Engineering, USSR): G-convergence and the eigenvalue problems.

A. MIKELIC (Rudjer Boskovic Institute, Zagreb, Yugoslavia): Homogenization of non stationary Navier-Stokes equations.

H.T. NGOAN (Academy of Sciences, Hanoi, Vietnam): Homogenization for operators in non-divergence form.

Contributors

Giovanni ALBERTI, Scuola Normale Superiore, Piazza dei Cavalieri 7, 56100 Pisa, Italy.

Marco AVELLANEDA, Courant Institute of Mathematical Sciences, 251 Mercer Street, New York, NY 10012, USA.

Nickolaj S. BAKHVALOV, Department of Numerical Mathematics, USSR Academy of Sciences, Leninskij Prospect 14, 117901 Moscow , USSR.

Alain BENSOUSSAN, Institut National de Recherche en Informatique et en Automatique (INRIA), Domaine de Voluceau, Rocquencourt, 78150 Le Chesnay, France.

David J. BERGMAN, School of Physics and Astronomy, Tel Aviv University, Sacler Faculty of Exact Science, Ramat Aviv, Tel Aviv 69978, Israel.

Lucio BOCCARDO, Dipartimento di Matematica, Università "La Sapienza", Piazzale A. Moro 2, 00185 Roma, Italy.

Guy BOUCHITTE, Mathématiques, Université de Toulon et du Var, Avenue de l'Université, BP 132, 83957 La Garde, France.

Giuseppe BUTTAZZO, Dipartimento di Matematica, Università di Ferrara, via Machiavelli 35, 44100 Ferrara, Italy.

Ennio DE GIORGI, Scuola Normale Superiore, Piazza dei Cavalieri 7, 56100 Pisa, Italy.

Margarita E. EGLIT, Department of Mathematics and Mechanics, Moscow State University, Moscow 119899, USSR.

Ulrich HORNUNG, Universität der Bundeswehr München, Fakultät für Informatik, Institut für Mathematik, P.O. Box 1222, Werner-Heisenberg Weg 39, 8014 Neubiberg , Federal Republic of Germany.

E. Ya. KHRUSLOV, Ucrainian SSR Academy of Sciences, Physico-Technical Institute of Low Temperatures, Lenin's Prospect 47, Kharkov 86, USSR.

Noboru KIKUCHI, Department of Mechanical Engineering and Applied Mechanics, The University of Michigan, Ann Arbor, MI 48109, USA.

S. M. KOZLOV, Moscow Institute of Civil Engineering, Moscow 115409, USSR.

Andrew J. MAJDA, Program for Computational and Applied Mathematics, Department of Mathematics, Princeton University, Princeron, NJ 08540, USA.

Graeme W. MILTON, Courant Institute of Mathematical Sciences, 251 Mercer Street, New York, NY 10012, USA.

Umberto MOSCO, Dipartimento di Matematica, Università "La Sapienza", Piazzale A. Moro 2, 00185 Roma, Italy.

François MURAT, Laboratoire d'Analyse Numérique, Université Paris VI, 4 place Jussieu, 75252 Paris, France.

Pierre SUQUET, Laboratoire de Méchanique et d'Acoustique, 31 Chemin Joseph-Aiguier, 13402 Marseille, France

Katsuyuki SUZUKI, Department of Mechanical Engineering and Applied Mechanics, The University of Michigan, Ann Arbor, MI 48109, USA.

Integral Representation of Functionals Defined on Sobolev Spaces

GIOVANNI ALBERTI GIUSEPPE BUTTAZZO

Abstract: We give an integral representation result for functionals defined on Sobolev spaces; more precisely, for a functional F, we find necessary and sufficient conditions that imply the integral representation formula

$$F(u,B) = \int_B f(x, Du)\, dx.$$

1. Introduction

The problem of representing in an integral form a given functional defined on a function space and satisfying suitable "abstract" conditions, has been considered by several authors in different frameworks (see References). One of the reasons is that it is the key point in many problems of relaxation and Γ-convergence (see for instance [3], [6], [8], [10], [11], [16], [20], [25]); in fact, the relaxed functionals (or the Γ-limits of a sequence of functionals) are merely lower semicontinuous mappings defined on a function space, and the first step in order to get their complete characterization, is just to represent them in a suitable integral form.

The most classical integral representation result is the well-known Riesz theorem which states that every linear continuous map $F : L^p(\Omega; \mathbf{R}^m) \to \mathbf{R}$ can be written in the form

$$F(u) = \int_\Omega f(x) \cdot u(x)\, dx \tag{1.1}$$

for a suitable $f \in L^q(\Omega;\mathbf{R}^m)$ (with $1/p + 1/q = 1$).

A nonlinear version of the Riesz representation theorem has been also proved (see for instance [14], [25], [28]); it states that every lower semicontinuous map $F : L^p(\Omega;\mathbf{R}^m) \to]-\infty,+\infty]$ which is disjointly additive in the sense that

$$F(u+v) = F(u) + F(v) \qquad \text{whenever } u \cdot v = 0 \text{ a.e. on } \Omega$$

can be represented in the form

$$(1.2) \qquad F(u) = \int_\Omega f\big(x, u(x)\big)\, dx$$

for a suitable Borel function $f(x,s)$ lower semicontinuous in s and such that

$$f(x,s) \geq -\big[a(x) + b|s|^p\big] \qquad \text{for all } (x,s) \in \Omega \times \mathbf{R}^m$$

with $a \in L^1(\Omega)$ and $b \geq 0$.

Other integral representation results for functionals defined on the space of measures, have also been proved (see [3], [6], [7], [8], [9], [10], [11], [20], [40]).

In this paper, we deal with functionals $F(u,B)$ defined for every u belonging to a Sobolev space $W^{1,p}(\Omega;\mathbf{R}^m)$ and every B belonging to the class $\mathcal{B}(\Omega)$ of all Borel subsets of Ω, and we look for an integral representation of F in the form

$$(1.3) \qquad F(u,B) = \int_B f\big(x, Du(x)\big)\, dx$$

for a suitable integrand $f(x,z)$. When F satisfies growth conditions as

$$(1.4) \qquad |F(u,B)| \leq \int_B \big[a(x) + b|Du|^p\big]\, dx$$

with $a \in L^1(\Omega)$ and $b \geq 0$, the integral representation formula (1.3) has been obtained by Buttazzo & Dal Maso in [15], [16] under the following additional hypotheses:

(i) F is local, that is

$$u = v \text{ a.e. on } B \in \mathcal{B}(\Omega) \;\Rightarrow\; F(u,B) = F(v,B)\,;$$

(ii) for every $u \in W^{1,p}(\Omega;\mathbf{R}^m)$ the set function $F(u,\cdot)$ is a measure on $\mathcal{B}(\Omega)$;

(iii) for every $u \in W^{1,p}(\Omega;\mathbf{R}^m)$, $c \in \mathbf{R}^m$, and $B \in \mathcal{B}(\Omega)$ we have

$$F(u+c,B) = F(u,B)\,;$$

(iv) for every $B \in \mathcal{B}(\Omega)$ the function $F(\cdot, B)$ is sequentially weakly lower semicontinuous on $W^{1,p}(\Omega; \mathbf{R}^m)$.

In this case the integrand $f(x, z)$ in (1.3) turns out to be quasi-convex with respect to z in the sense of Morrey [36].

Here we follow a different approach based on a recent result by Alberti (see [1]) concerning a Lusin type property for L^p-functions (Theorem 2.7). This will enable us to obtain the integral representation (1.3) even if the growth condition (1.4) is dropped and condition (iv) is substituted by the weaker one:

(iv') for every $B \in \mathcal{B}(\Omega)$ the function $F(\cdot, B)$ is lower semicontinuous on $W^{1,p}(\Omega; \mathbf{R}^m)$ with respect to the strong topology.

2. Notation and Statement of the Result

In this section we fix the notation we shall use in the following and we state our main result. We also recall some other results which will be used in the proofs.

Let Ω be a bounded open subset of $\mathbf{R}^n$, let $m \geq 1$ be an integer, and let $p \in [1, +\infty]$; we denote by $W^{1,p}(\Omega; \mathbf{R}^m)$ the usual Sobolev space with norm

$$||u||_{W^{1,p}(\Omega;\mathbf{R}^m)} = ||u||_{L^p(\Omega;\mathbf{R}^m)} + ||Du||_{L^p(\Omega;\mathbf{R}^{mn})}.$$

For every $u \in W^{1,p}(\Omega; \mathbf{R}^m)$ and for a.e. $x \in \Omega$ the gradient $Du(x)$ will be the $m \times n$ matrix defined by $\big(Du(x)\big)_{i,j} = D_j u_i(x)$ for $i = 1, \dots, m$ and $j = 1, \dots, n$.

We shall consider functionals $F : W^{1,p}(\Omega; \mathbf{R}^m) \times \mathcal{B}(\Omega) \to [0, +\infty]$ where $\mathcal{B}(\Omega)$ denotes the class of all Borel subsets of Ω. For this kind of functionals we introduce the following definitions.

Definition 2.1. *We say that a functional $F : W^{1,p}(\Omega; \mathbf{R}^m) \times \mathcal{B}(\Omega) \to [0, +\infty]$ is*

(i) local, if $F(u, B) = F(v, B)$ whenever $B \in \mathcal{B}(\Omega)$ and $u, v \in W^{1,p}(\Omega; \mathbf{R}^m)$ with $u = v$ a.e. on B;

(ii) D-local, if $F(u, B) = F(v, B)$ whenever $B \in \mathcal{B}(\Omega)$ and $u, v \in W^{1,p}(\Omega; \mathbf{R}^m)$ with $Du = Dv$ a.e. on B;

(iii) a measure, if for every $u \in W^{1,p}(\Omega; \mathbf{R}^m)$ the set function $F(u, \cdot)$ is countably additive on $\mathcal{B}(\Omega)$.

We are now in a position to state our integral representation result.

Theorem 2.2. *Let $p \in [1, +\infty[$, and let $F : W^{1,p}(\Omega; \mathbf{R}^m) \times \mathcal{B}(\Omega) \to [0, +\infty]$ be a functional such that:*

(i) F is D-local;

(ii) F *is a measure;*

(iii) *for every* $B \in \mathcal{B}(\Omega)$ *the function* $F(\cdot, B)$ *is lower semicontinuous on* $W^{1,p}(\Omega; \mathbf{R}^m)$ *with respect to the strong topology;*

(iv) *there exists* $\bar{u} \in W^{1,p}(\Omega; \mathbf{R}^m)$ *such that* $F(\bar{u}, \cdot)$ *is a bounded measure which is absolutely continuous with respect to the Lebesgue measure.*

Then there exists a Borel function $f : \Omega \times \mathbf{R}^{mn} \to [0, +\infty]$ *such that*

(a) *for every* $x \in \Omega$ *the function* $f(x, \cdot)$ *is lower semicontinuous on* $\mathbf{R}^{mn}$*;*

(b) *for every* $(u, B) \in W^{1,p}(\Omega; \mathbf{R}^m) \times \mathcal{B}(\Omega)$ *it is*

$$F(u, B) = \int_B f(x, Du(x))\, dx\,.$$

Moreover, the integrand f *is uniquely determined in the following sense: if* g *is a Borel function so that (b) holds with* g *instead of* f*, then there exists a negligible set* $N \in \mathcal{B}(\Omega)$ *such that* $f(x, s) = g(x, s)$ *for all* $x \in \Omega \setminus N$ *and* $s \in \mathbf{R}^{mn}$*.*

Remark 2.3. Note that hypotheses (i) and (iv) of Theorem 2.2 yield $F(u, B) = F(\bar{u}, B) = 0$ for all $u \in W^{1,p}(\Omega; \mathbf{R}^m)$ and all $B \in \mathcal{B}(\Omega)$ with $|B| = 0$.

Remark 2.4. For simplicity we consider only the case $p < +\infty$; in the case $p = +\infty$ the same result (with the same proof) holds, provided condition (iii) is substituted by the following one:

(iii') *for every* $B \in \mathcal{B}(\Omega)$ *the function* $F(\cdot, B)$ *is lower semicontinuous on* $W^{1,\infty}(\Omega; \mathbf{R}^m)$ *with respect to the* τ_∞*-convergence, where we say that* u_h *is* τ_∞*-convergent to* u *if* u_h *is bounded in* $W^{1,\infty}(\Omega; \mathbf{R}^m)$*,* u_h *converges to* u *uniformly on compact subsets of* Ω*, and* Du_h *converges to* Du *a.e. in* Ω*.*

Remark 2.5. The hypothesis that F is positive can be easily weakened by requiring that for suitable $a \in L^1(\Omega)$ and $b \geq 0$

$$F(u, B) \geq - \int_B \left[a(x) + b|Du|^p\right] dx$$

for all $(u, B) \in W^{1,p}(\Omega; \mathbf{R}^m) \times \mathcal{B}(\Omega)$.

Remark 2.6. By Definition 2.1 and by the well-known locality of the gradient (see for instance Gilbarg & Trudinger [27], Lemma 7.7), it follows immediately that D-locality implies locality. The converse implication can be also proved (we refer to Alberti [2] for the proof) provided F satisfies condition (ii) of Theorem 2.2 and the invariance condition (iii) stated in the Introduction. Finally, when F satisfies conditions (ii) and (iii) of Theorem 2.2 and the growth condition (1.4), then the locality of F follows from the locality on open sets (see Buttazzo & Dal Maso [15], Lemma 2.8), that is

$F(u,A) = F(v,A)$ whenever A is an open subset of Ω, and $u, v \in W^{1,p}(\Omega;\mathbf{R}^m)$ with $u = v$ a.e. on A.

The main tools in the proof of the integral representation Theorem 2.2 are the following results.

Theorem 2.7. (See Alberti [1], Theorem 1) *For every $v \in L^p(\Omega;\mathbf{R}^{mn})$ and every $\varepsilon > 0$ there exist a function $u \in C_0^1(\Omega;\mathbf{R}^m)$ and a closed set $B \subset \Omega$ such that*

(i) $|\Omega \setminus B| \leq \varepsilon|\Omega|$;
(ii) $v = Du$ a.e. in B;
(iii) $||Du||_p \leq C\varepsilon^{1/p-1}||f||_p$ where C is a constant which depends only on n.

Theorem 2.8. (See Buttazzo & Dal Maso [14]) *Let $m \geq 1$ be a given integer and let $F : L^p(\Omega;\mathbf{R}^m) \times \mathcal{B}(\Omega) \to [0,+\infty]$ be a functional such that*

(i) for all $v \in L^p(\Omega;\mathbf{R}^m)$, the function $F(v,\cdot)$ is a measure which is absolutely continuous with respect to the Lebesgue measure;
(ii) F is local, that is $F(v,B) = F(v',B)$ whenever $v = v'$ a.e. in B;
(iii) for all $B \in \mathcal{B}(\Omega)$ the function $F(\cdot,B)$ is lower semicontinuous with respect to strong topology of $L^p(\Omega;\mathbf{R}^m)$;
(iv) there exists $\bar{v} \in L^p(\Omega;\mathbf{R}^m)$ such that $F(\bar{v},\Omega) < +\infty$.

Then there exists a Borel function $f : \Omega \times \mathbf{R}^m \to [0,+\infty]$ such that

(a) for every $x \in \Omega$ the function $f(x,\cdot)$ is lower semicontinuous on $\mathbf{R}^m$;
(b) for every $(v,B) \in L^p(\Omega;\mathbf{R}^m) \times \mathcal{B}(\Omega)$ it is

$$F(v,B) = \int_B f\big(x, v(x)\big)\, dx\,.$$

3. Proof of the Result

For the sake of simplicity we always refer to $W^{1,p}(\Omega;\mathbf{R}^m)$ as $W^{1,p}$, to $C_0^1(\Omega;\mathbf{R}^m)$ as C_0^1 and to $L^p(\Omega;\mathbf{R}^{mn})$ as L^p.

Let F be a functional on $W^{1,p}$ satisfying conditions (i), (ii), (iii), (iv) of Theorem 2.2. We shall apply Theorem 2.7 to find a functional G on L^p which satisfies hypotheses (i), (ii), (iii), (iv) of Theorem 2.8 and such that

$$F(u,B) = G(Du,B) \qquad \text{for all } (u,B) \in W^{1,p} \times \mathcal{B}(\Omega)\,.$$

Definition 3.1. *Let $v \in L^p$ and let $(u_h, B_h) \in W^{1,p} \times \mathcal{B}(\Omega)$ for every $h \in \mathbf{N}$. We say that (u_h, B_h) is a local partition of v if*

(3.1) the sets B_h are pairwise disjoint and cover almost all of Ω;
(3.2) $Du_h = v$ a.e. in B_h for every $h \in \mathbf{N}$.

Proposition 3.2. *For every $v \in L^p$ and every $\varepsilon > 0$ there exists a local partition (u_h, B_h) of v such that*

$$|\Omega \setminus B_0| < \varepsilon \qquad \text{and} \qquad \|u_0\|_{W^{1,p}} \leq C\varepsilon^{1/p-1}\|v\|_p$$

where C is a constant which does not depend on v.

Proof. Fix $v \in L^p$ and $\varepsilon > 0$. By Theorem 2.7 there exist functions $u_h \in C_0^1$ and closed sets $A_h \subset \Omega$ such that
(i) $|\Omega \setminus A_h| < \varepsilon 2^{-h}$ for every $h \in \mathbf{N}$;
(ii) $Du_h = v$ a.e. in A_h;
(iii) $\|Du_h\|_p \leq C\left(\varepsilon|\Omega|^{-1}2^{-h}\right)^{1/p-1}\|v\|_p$, where C is the constant of Theorem 2.7.

Setting for every $h \in \mathbf{N}$

$$B_h = A_h \setminus \bigcup_{j=0}^{h-1} A_j$$

it is easy to verify that $|\Omega \setminus \bigcup_0^\infty B_h| = 0$. Moreover, by Poincaré inequality, we get

$$\|u_0\|_{W^{1,p}} \leq C'\|Du_0\|_p \leq C'C|\Omega|^{1-1/p}\varepsilon^{1/p-1}\|v\|_p$$

where C' is a constant which does not depends on v. Hence Proposition 3.2 is proved. ■

Proposition 3.3. *Let $v \in L^p$ and let (u_h, A_h) and (u'_h, A'_h) be two local partitions of v. Then it is*

$$F(u_h, B) = F(u'_k, B)$$

for all $h, k \in \mathbf{N}$ and all Borel sets $B \subset A_h \cap A'_k$. In particular, for all $B \in \mathcal{B}(\Omega)$ we have

$$\sum_{h\in\mathbf{N}} F(u_h, B \cap A_h) = \sum_{k\in\mathbf{N}} F(u'_k, B \cap A'_k)\,. \tag{3.3}$$

Proof. Since $v = Du_h = Du'_k$ a.e in $A_h \cap A'_k$, by hypothesis (i) of Theorem 2.2 we obtain that $F(u_h, B) = F(u'_k, B)$ for all integers h, k and all Borel sets $B \subset A_h \cap A'_k$. Taking into account (3.1) and Remark 2.3, this yields

$$\sum_{h\in\mathbf{N}} F(u_h, B \cap A_h) = \sum_{h,k\in\mathbf{N}} F(u_h, B \cap A_h \cap A'_k)$$
$$= \sum_{h,k\in\mathbf{N}} F(u'_k, B \cap A_h \cap A'_k) = \sum_{k\in\mathbf{N}} F(u'_k, B \cap A'_k)\,. \blacksquare$$

Lemma 3.4. *For all* $(v, B) \in L^p \times \mathcal{B}(\Omega)$ *define*

$$G(v, B) = \sum_{h \in \mathbf{N}} F(u_h, B \cap A_h) \tag{3.4}$$

where (u_h, A_h) *is a local partition of* v. *We have:*

(i) G *is well-defined, in the sense that* $G(v, B)$ *does not depend on the choice of the local partition of* v;

(ii) *for all* $(u, B) \in W^{1,p} \times \mathcal{B}(\Omega)$ *it is* $F(u, B) = G(Du, B)$.

Proof. The fact that G is well-defined follows from Proposition 3.3. In order to prove (ii), set $(u_0, A_0) = (u, \Omega)$ and $(u_h, A_h) = (0, \emptyset)$ for $h \geq 1$ and note that this is a local partition for Du. By the definition of G we have

$$G(Du, B) = \sum_{h \in \mathbf{N}} F(u_h, B \cap A_h) = F(u, B) \qquad \text{for all } B \in \mathcal{B}(\Omega). \blacksquare$$

Lemma 3.5. *For all* $v \in W^{1,p}$ *the function* $G(v, \cdot)$ *is a positive measure which is absolutely continuous with respect to the Lebesgue measure.*

Proof. Let v be a function in L^p, let $B \in \mathcal{B}(\Omega)$, and let $(B_k)_{k \in \mathbf{N}}$ be a partition of B into Borel sets. If (u_h, A_h) is a local partition of v, by using the definition of G, hypothesis (i) of Theorem 2.2, and Remark 2.3,we get

$$\begin{aligned}\sum_{k \in \mathbf{N}} G(v, B_k) &= \sum_{k \in \mathbf{N}} \Big[\sum_{h \in \mathbf{N}} F(u_h, B_k \cap A_h) \Big] \\ &= \sum_{h \in \mathbf{N}} \Big[\sum_{k \in \mathbf{N}} F(u_h, B_k \cap A_h) \Big] \\ &= \sum_{h \in \mathbf{N}} F(u_h, B \cap A_h) = G(v, B) .\end{aligned}$$

Therefore $G(v, \cdot)$ is a measure. The fact that $G(v, B) = 0$ whenever $|B| = 0$ is obvious. ■

Lemma 3.6. G *is local, that is* $G(v, B) = G(v', B)$ *for all* $v, v' \in L^p$ *and all* $B \in \mathcal{B}(\Omega)$ *such that* $v = v'$ *a.e in* B.

Proof. Let $v, v' \in L^p$ and let $B \in \mathcal{B}(\Omega)$ such that $v = v'$ a.e in B. If (u_h, A_h) and (u'_h, A'_h) are local partitions of v and v' respectively, by (3.2) we get $Du_h = Du'_k$ a.e. in $B \cap A_h \cap A'_k$ for all integers h, k. Hence, taking into account hypothesis (ii) of Theorem 2.2,

$$F(u_h, B \cap A_h \cap A'_k) = F(u'_k, B \cap A_h \cap A'_k)$$

for all $h, k \in \mathbf{N}$. Arguing as in the proof of Proposition 3.3 we obtain

$$\begin{aligned} G(v,B) &= \sum_{h\in\mathbf{N}} F(u_h, B\cap A_h) \\ &= \sum_{h,k\in\mathbf{N}} F(u_h, B\cap A_h\cap A'_k) = \sum_{h,k\in\mathbf{N}} F(u'_k, B\cap A_h\cap A'_k) \\ &= \sum_{k\in\mathbf{N}} F(u'_k, B\cap A'_k) = G(v',B)\ .\ \blacksquare \end{aligned}$$

Lemma 3.7. *For all $B \in \mathcal{B}(\Omega)$ the function $G(\cdot, B)$ is lower semicontinuous in the strong topology of L^p.*

Proof. An easy computation shows that it is enough to prove that

$$G(v,B) \le \liminf_{h\to\infty} G(v+v_h, B) \tag{3.5}$$

whenever v_h are functions in L^p such that $\|v_h\|_p \le 4^{-h}$ for every $h \in \mathbf{N}$. By Proposition 3.2, for every $h \in \mathbf{N}$ we may choose a local partition $(u_{h,j}, A_{h,j})$ of v_h such that

$$|\Omega \setminus A_{h,0}| < 2^{-h} \qquad \text{and} \qquad \|u_{h,0}\|_{W^{1,p}} \le C2^{(1-1/p)h}\|v_h\|_p \le C2^{-h}\,, \tag{3.6}$$

where C is a constant which does not depend on h.
Fix an integer k. Choose a local partition (u_h, A_h) of v such that $|\Omega\setminus A_0| \le 2^{-k}$ (cf. Proposition 3.2) and set

$$C_k = A_0 \cap \Big(\bigcap_{h\ge k} A_{h,0}\Big)\ .$$

By the definition of G, for all $h, k \in \mathbf{N}$ we get

$$\begin{aligned} &G(v, B\cap C_k) = F(u_0, B\cap C_k) \\ &G(v+v_h, B\cap C_k) = F(u_0+u_{h,0}, B\cap C_k) \qquad \text{whenever}\, h \ge k, \end{aligned}$$

and taking into account that $u_0 + u_{h,0}$ converge to u_0 by (3.6), and that F is lower semicontinuous (hypothesis (iii) of Theorem 2.2), we obtain

$$\begin{aligned} G(v, B\cap C_k) &= F(u_0, B\cap C_k) \\ &\le \liminf_{h\to\infty} F(u_0+u_{h,0}, B\cap C_k) \\ &= \liminf_{h\to\infty} G(v_0+v_{h,0}, B\cap C_k)\ . \end{aligned} \tag{3.7}$$

Note that by definition of C_k and by (3.6)

$$|\Omega \setminus C_k| \leq |\Omega \setminus A_0| + \sum_{h \geq k} |\Omega \setminus A_{h,0}| \leq 2^{-k} + \sum_{h \geq k} 2^{-h} = 3 \cdot 2^{-k},$$

so that $|\Omega \setminus C_k|$ converge to 0 as $k \to \infty$. Hence, for every $t < G(v, B)$ there exists an integer k such that $t \leq G(v, B \cap C_k)$ and inequality (3.7) and the fact that G is positive yield

$$\begin{aligned} t &\leq G(v, B \cap C_k) \\ &\leq \liminf_{h \to \infty} G\left(v_0 + v_{h,0}, B \cap C_k\right) \\ &\leq \liminf_{h \to \infty} G\left(v_0 + v_{h,0}, B\right) \end{aligned}$$

Therefore (3.5) is satisfied because t is any real number less than $G(v, B)$. ∎

Proof of Theorem 2.2. By Lemmas 3.5, 3.6 and 3.7 we have that G satisfies hypotheses (i), (ii), (iii) of Theorem 2.8. Lemma 3.4 and (iv) of Theorem 2.2 imply that hypothesis (iv) of Theorem 2.8 holds with $\bar{v} = D\bar{u}$. Then there exists a Borel function $f : \Omega \times \mathbf{R}^{mn} \to [0, +\infty]$ which is lower semicontinuous in the second variable and such that

$$G(v, B) = \int_B f\left(x, v(x)\right) dx$$

for every $(v, B) \in L^p \times \mathcal{B}(\Omega)$. Lemma 3.4 again implies

$$F(u, B) = G(Du, B) = \int_B f\left(x, Du(x)\right) dx$$

for every $(u, B) \in W^{1,p} \times \mathcal{B}(\Omega)$, and then we have proved (a) and (b). The uniqueness of the integrand f follows for instance from Corollary 6 of Alberti [1]. ∎

References

[1] G.ALBERTI: *A Lusin type theorem for gradients*. Preprint Scuola Normale Superiore, Pisa (1990).

[2] G.ALBERTI: Paper in preparation.

[3] L.AMBROSIO & G.BUTTAZZO: *Weak lower semicontinuous envelope of functionals defined on a space of measures*. Ann. Mat. Pura Appl., **150** (1988), 311–340.

[4] J.APPELL: *The Superposition Operator in Function Spaces. A Survey.* Book in preparation.

[5] G.BOTTARO & P.OPPEZZI: *Rappresentazione con integrali multipli di funzionali dipendenti da funzioni a valori in uno spazio di Banach.* Ann. Mat. Pura Appl., **139** (1985), 191–225.

[6] G.BOUCHITTE: *Représentation intégrale de fonctionnelles convexes sur un espace de mesures.* Ann. Univ. Ferrara, **33** (1987), 113–156.

[7] G.BOUCHITTE & G.BUTTAZZO: *New lower semicontinuity results for non convex functionals defined on measures.* Nonlinear Anal., (to appear).

[8] G.BOUCHITTE & G.BUTTAZZO: *Non convex functionals defined on measures: integral representation and relaxation.* Paper in preparation.

[9] G.BOUCHITTE & M.VALADIER: *Integral representation of convex functionals on a space of measures.* J. Funct. Anal., **80** (1988), 398–420.

[10] G.BOUCHITTE & M.VALADIER: *Multifonctions s.c.i. et régularisée s.c.i. essentielle. Fonctions de mesure dans le cas sous linéaire.* Proceedings "Congrès Franco-Québécois d'Analyse Non Linéaire Appliquée", Perpignan, June 22–26, 1987, Bordas, Paris (1989).

[11] G.BUTTAZZO: *Semicontinuity, Relaxation and Integral Representation in the Calculus of Variations.* Pitman Res. Notes Math. Ser. **207**, Longman, Harlow (1989).

[12] G.BUTTAZZO: *Semicontinuity, relaxation, and integral representation problems in the calculus of variations.* Notes of a series of lectures held at CMAF of Lisbon in November–December 1985. Printed by CMAF, Lisbon (1986).

[13] G.BUTTAZZO & G.DAL MASO: *Integral representation on $W^{1,\alpha}(\Omega)$ and $BV(\Omega)$ of limits of variational integrals.* Atti Accad. Naz. Lincei Rend. Cl. Sci. Fis. Mat. Natur., **66** (1979), 338–343.

[14] G.BUTTAZZO & G.DAL MASO: *On Nemyckii operators and integral representation of local functionals.* Rend. Mat., **3** (1983), 491–509.

[15] G.BUTTAZZO & G.DAL MASO: *A characterization of nonlinear functionals on Sobolev spaces which admit an integral representation with a Carathéodory integrand.* J. Math. Pures Appl., **64** (1985), 337–361.

[16] G.BUTTAZZO & G.DAL MASO: *Integral representation and relaxation of local functionals.* Nonlinear Anal., **9** (1985), 512–532.

[17] G.DAL MASO: *Integral representation on $BV(\Omega)$ of Γ-limits of variational integrals.* Manuscripta Math., **30** (1980), 387–413.

[18] G.DAL MASO: *On the integral representation of certain local functionals.* Ricerche Mat., **32** (1983), 85–131.

[19] G.DAL MASO & L.MODICA: *A general theory of variational integrals.* Quaderno della Scuola Normale Superiore "Topics in Functional Analysis 1980–81", Pisa (1982), 149–221.

[20] E.DE GIORGI & L.AMBROSIO & G.BUTTAZZO: *Integral representation and relaxation for functionals defined on measures.* Atti Accad. Naz. Lincei Rend. Cl. Sci. Fis. Mat. Natur., **81** (1987), 7–13.

[21] L.DREWNOWSKI & W.ORLICZ: *On orthogonally additive functionals.* Bull. Polish Acad. Sci. Math., **16** (1968), 883–888.

[22] L.DREWNOWSKI & W.ORLICZ: *Continuity and representation of orthogonally additive functionals.* Bull. Polish Acad. Sci.Math., **17** (1969), 647–653.

[23] F.FERRO: *Integral characterization of functionals defined on spaces of BV functions.* Rend. Sem. Mat. Univ. Padova, **61** (1979), 177–203.

[24] A.FOUGERES & A.TRUFFERT: *Δ-integrands and essential infimum, Nemyckii representation of l.s.c. operators on decomposable spaces and Radon-Nikodym-Hiai representation of measure functionals.* Preprint A.V.A.M.A.C. University of Perpignan, Perpignan (1984).

[25] A.FOUGERES & A.TRUFFERT: *Applications des méthodes de représentation intégrale et d'approximation inf-convolutives à l'épi-convergence.* Preprint A.V.A.M.A.C. University of Perpignan, Perpignan (1985).

[26] N.FRIEDMAN & M.KATZ: *Additive functionals of L^p spaces.* Canad. J. Math., **18** (1966), 1264–1271.

[27] D.GILBARG & N.S.TRUDINGER: *Elliptic Partial Differential Equations of Second Order.* Springer-Verlag, Berlin (1977).

[28] F.HIAI: *Representation of additive functionals on vector valued normed Kothe spaces.* Kodai Math. J., **2** (1979), 300–313.

[29] M.MARCUS & V.J.MIZEL: *Nemyckii operators on Sobolev spaces.* Arch. Rational Mech. Anal., **51** (1973), 347–370.

[30] M.MARCUS & V.J.MIZEL: *Extension theorems for nonlinear disjointly additive functionals and operators on Lebesgue spaces, with applications.* Bull. Amer. Math. Soc., **82** (1976), 115–117.

[31] M.MARCUS & V.J.MIZEL: *Extension theorems of Hahn-Banach type for nonlinear disjointly additive functionals and operators in Lebesgue spaces.* J. Funct. Anal., **24** (1977), 303–335.

[32] M.MARCUS & V.J.MIZEL: *Representation theorems for nonlinear disjointly additive functionals and operators on Sobolev spaces.* Trans. Amer. Math. Soc., **228** (1977), 1–45.

[33] M.MARCUS & V.J.MIZEL: *A characterization of first order nonlinear partial differential operators on Sobolev spaces.* J. Funct. Anal., **38** (1980), 118–138.

[34] V.J.MIZEL: *Characterization of nonlinear transformations possessing kernels.* Canad. J. Math., **22** (1970), 449–471.

[35] V.J.MIZEL & K.SUNDARESAN: *Representation of vector valued nonlinear functions.* Trans. Amer. Math. Soc., **159** (1971), 111–127.

[36] C.B.MORREY: *Quasiconvexity and the semicontinuity of multiple integrals.* Pacific J. Math., **2** (1952), 25–53.

[37] C.SBORDONE: *Sulla caratterizzazione degli operatori differenziali del* 2° *ordine di tipo ellittico.* Rend. Accad. Sci. Fis. Mat. Napoli, **41** (1975), 31–45.

[38] S.SPAGNOLO: *Una caratterizzazione degli operatori differenziali autoaggiunti del* 2° *ordine a coefficienti misurabili e limitati.* Rend. Sem. Mat. Univ. Padova, **38** (1967), 238–257.

[39] I.V.SRAGIN: *Abstract Nemyckii operators are locally defined operators.* Soviet Math. Dokl., **17** (1976), 354–357.

[40] M.VALADIER: *Fonctions et opérateurs sur les mesures.* C. R. Acad. Sci. Paris, **I-304** (1987), 135–137.

[41] W.A.WOYCZYNSKI: *Additive functionals on Orlicz spaces.* Colloq. Math., **19** (1968), 319–326.

Giovanni Alberti

Scuola Normale Superiore
Piazza dei Cavalieri, 7
56126 PISA (ITALY)

Giuseppe Buttazzo

Dipartimento di Matematica
Via Machiavelli, 35
44100 FERRARA (ITALY)

HOMOGENIZATION AND RENORMALIZATION OF MULTIPLE-SCATTERING EXPANSIONS FOR GREEN FUNCTIONS IN TURBULENT TRANSPORT

by

Marco AVELLANEDA
and
Andrew J. MAJDA

1. Introduction

The purpose of this article is to report on a method for averaging equations with rapidly varying characteristics, based on the asymptotic analysis of the perturbation expansions for the corresponding Green functions. While such methodology is relevant to many problems in homogenization, we shall discuss it primarily in the context of advection-diffusion equations describing passive turbulence transport.

Perturbative techniques for the study of Green functions corresponding to various processes in heterogeneous or disordered media have been extensively used, both theoretically and computationally. The use of such expansions, known as multiple-scattering expansions, goes back at least to the work of M. Lax[1]. We refer also to the work of Bixon and Zwanzig[2] and Kirpatrick[3], that use perturbative methods to study transport properties in composite materials formed of a homogeneous matrix with inclusions ; see also Fredrickson and Shaqfeh[4] and references therein. Early work on perturbation methods is discussed in Beran[5] and various developments (until the late 70's), especially in relation to the overall properties of composites, are reviewed in Willis[6].

More recently, a connection was established between the multiple-scattering expansion for the effective properties of composites, and complex function theory. This development originated in the work of Bergman[7], followed by Milton[8] and Golden and Papanicolaou[9]. These authors recognized that the effective physical constants of two-phase composites depend analytically on the perturbation parameter measuring the contrast between the properties of the two media. In fact, they showed that the effective conductivity admits a Stieltjes integral representation which provides naturally an analytic continuation for the multiple-scattering expansion to the entire complex plane minus a set contained in a segment or half-line, corresponding to unphysical values of the perturbation parameter.

A similar integral representation formula was obtained by Avellaneda and Majda[10,11] for the effective diffusion coefficient for the problem of passive advection (with diffusion) of a scalar quantity by a time-independent, incompressible velocity field. Here, the perturbation parameter was taken to be the Péclet number, which measures the strength of the coupling between diffusion and advection. Previously, Wolynes[12] derived a similar integral representation formula for the effective diffusivity of mode-coupling systems near equilibrium.

The importance of these integral representations from a theoretical viewpoint, is that they provide resummation formulas for the multiple-scattering expansion at large (but physical) values of the perturbation parameter. In particular, the method of Padé approximants can then be used to obtain bounds on the effective constants. Such ideas were also developed by several authors to estimate the matrix elements of Green functions of quantum systems[13], and by Kraichnan[14], to study long-time properties of turbulent transport. In other developments, Tartar[15], and later Amirat, Hamdache and Ziani[16,17], combined perturbation theory of linear operators, function-theoretic arguments and the method of Young measures, to derive effective equations for certain evolution equations. In some cases, the effective equations that they obtained were non-local. In this paper, we shall use the analytic continuation properties of the perturbation equation for the resolvent operator and, in particular, integral representations for the averaged Green function, as the main tools for carrying out the homogenization.

To show the perturbation method at work, we shall discuss an elementary but interesting example : a parabolic advection-diffusion equation with a quasiperiodic or random first-order advective term. Due to the special form of the equation -the advective field is assumed to be incompressible and time-independent- we obtain a simple integral representation for the averaged

Green function.

Our choice of the advection-diffusion equation is also motivated by the fact that it exhibits different scaling laws and effective Green functions, according to the singularity of the Fourier transform of the advective field near $\mathbf{k} = 0$. The crossover between normal and anomalous diffusion that ocurrs for this equation is discussed in Mathéron and de Marsily[18], Bouchaud, Comptet, Georges and Le Doussal[19], Koch and Brady[20] and Avellaneda and Majda[21]. It is caused in part by the fact that the equation contains infinitely many excited scales of motion. The perturbation method provides in our opinion interesting insight on this difficult problem.

The exposition is organized as follows : in Section 2, we develop the basic perturbation theory for the averaged Green function of advection-diffusion equations. In Section 3, we derive the renormalized representation formula, following Kraichnan[14]. In Section 4, we apply these results to obtain a homogenization theorem under the mean-field conditions on the advective term $\mathbf{u}$. This theorem was established by Oelshlager[22] using probabistic methods and recently by Avellaneda and Majda[11], using the framework of homogenization of equations with random coefficients developed by Papanicolaou and Varadhan[23]. The perturbation approach also yields in a very natural way a Stieltjes integral representation for the effective diffusivity obtained by Avellaneda and Majda in Refs 10 and 11. Finally, Section 5 is devoted to the study of the scaling laws and effective Green functions for a class of examples consisting of quasi-periodic simple shear flows. A rather precise form of the averaged Green function, which can be interpreted as having a non-local, frequency/wave number dependent effective diffusivity is derived. The method can be generalized to handle various kinds of quasi periodic or random simple shear flows. These calculations complement -using an entirely different method- the recent results of Avellaneda and Majda[21] for Gaussian flows, obtained using a probabilistic representation formula.

The examples discussed here by no means exhaust the range of application of the perturbation method in homogenization. For instance, this approach seems particularly useful to analyze the homogenization of problems involving pseudo-differential operators. Hopefully, such methods, combined with other homogenization techniques, will be useful to study some of the more challenging open problems in homogenization theory involving anomalous scaling laws and/or non-local effective equations.

2. Perturbation theory for advection-diffusion equations

We consider the equation

$$\frac{\partial}{\partial t} T(\mathbf{x},t) + z\mathbf{u}(\mathbf{x}) \cdot \nabla T(\mathbf{x},t) = D_0\, \Delta T(\mathbf{x},t) \tag{1}$$

where $x \in \mathbb{R}^d$, $d \geq 2$, D_0 is the diffusion coefficient and z is a non-dimensional coupling parameter. We assume that $\mathbf{u}(\mathbf{x})$ is a stationary, or statistically homogeneous stochastic process, with zero mean, satisfying the incompressibility condition

$$\nabla \cdot \mathbf{u}(\mathbf{x}) = 0 \quad , \quad \mathbf{x} \in \mathbb{R}^d. \tag{2}$$

To fix ideas, we shall consider fields $\mathbf{u}(\mathbf{x})$ of the form

$$\mathbf{u}(\mathbf{x}) = \sum_{\substack{\mathbf{k}\neq 0 \\ |\mathbf{k}|\leq\Lambda}} e^{i\mathbf{k}\cdot\mathbf{x}}\, \hat{\mathbf{u}}(\mathbf{k}) \tag{3}$$

where the sum ranges over a countable, bounded set in reciprocal space. The number Λ is a high-wavenumber cutoff that corresponds in certain physical settings to the reciprocal of the dissipation length scale. The "modes" $\hat{\mathbf{u}}(\mathbf{k})$ in (3) will be either non-random, in which case $\mathbf{u}(\mathbf{x})$ is quasi-periodic, or random. In the latter case we assume that $\mathbf{u}(\mathbf{x})$ is ergodic under spatial translations, and that

$$< \hat{\mathbf{u}}(\mathbf{k}) >= 0 \quad , \text{ for all } \mathbf{k} \neq 0, \tag{5}$$

and

$$< \hat{\mathbf{u}}_\alpha(\mathbf{k})\hat{\mathbf{u}}_\beta(\mathbf{k}') >= \delta(\mathbf{k}+\mathbf{k}')\hat{R}_{\alpha\beta}(\mathbf{k}), \tag{6}$$

for some tensor-valued function $\hat{R}_{\alpha\beta}$. In (5), (6) brackets denote statistical or infinite volume (Bohr) averaging. The tensor $\hat{R}_{\alpha\beta}(\mathbf{k})$ is the Fourier transform of the two-point correlation function of $\mathbf{u}(\mathbf{x})$ since, from (3),

$$< \mathbf{u}_\alpha(\mathbf{x})\mathbf{u}_\beta(\mathbf{y}) >= \sum_{\mathbf{k}} e^{i\mathbf{k}\cdot(\mathbf{x}-\mathbf{y})}\hat{R}_{\alpha\beta}(\mathbf{k}). \tag{7}$$

We see from this formula that the behavior of $\hat{R}_{\alpha\beta}(\mathbf{k})$ for $|\mathbf{k}| \ll 1$ is related to the decay of the statistical correlations of the field at large distances. It is well-known that the existence of long-range statistical correlations leads to the phenomenon of anomalous diffusion and that the long-time/large-distance behavior of (1) is of radically different nature according to the behavior of $\hat{R}(\mathbf{k})$ for $|\mathbf{k}| \ll 1$.[18,19,20,21].

To study the properties of solutions of (1) we consider as a starting point the operators

$$\mathbf{P}_z(t) = \exp t\{D_0\,\Delta - z\mathbf{u}(x)\cdot\nabla\} \tag{8}$$

and

$$\begin{aligned}\mathbf{G}_z(\lambda) &= \int_0^\infty e^{-\lambda t}\,\mathbf{P}_z(t)dt \\ &= (\lambda - D_0\,\Delta + z\mathbf{u}\cdot\nabla)^{-1}\end{aligned} \tag{9}$$

together with their respective kernels, $P_z(t,\mathbf{x},\mathbf{y})$ and $G_z(\lambda,\mathbf{x},\mathbf{y})$. For $z=0$, these kernels and translation invariant, with Fourier transforms given by

$$\overline{P}_0(t;\boldsymbol{\xi}) = e^{-tD_0|\boldsymbol{\xi}|^2}\ ,\ \overline{G}_0(\lambda,\boldsymbol{\xi}) = \frac{1}{\lambda + D_0|\boldsymbol{\xi}|^2}. \tag{10}$$

Although $P_z(t,\mathbf{x},\mathbf{y})$ and $G_z(\lambda,\mathbf{x},\mathbf{y})$ are not translation invariant for $z\neq 0$, the averaged functions $<\mathbf{P}_z(t)\psi>, <\mathbf{G}_z(\lambda)\psi>$, where $\psi\in C_0^\infty(\mathbb{R}^d)$, are also given by Fourier multipliers : i.e., we have

$$<\mathbf{P}_z(t)\psi> = \int_{\mathbb{R}^d} \overline{P}_z(t,\boldsymbol{\xi})e^{i\mathbf{x}\cdot\boldsymbol{\xi}}\,\hat{\psi}(\boldsymbol{\xi})d\boldsymbol{\xi} \tag{11}$$

and

$$<\mathbf{G}_z(\lambda)\psi> = \int_{\mathbb{R}^d} \overline{G}_z(\lambda,\boldsymbol{\xi})e^{i\mathbf{x}\cdot\boldsymbol{\xi}}\,\hat{\psi}(\boldsymbol{\xi})d\boldsymbol{\xi}, \tag{12}$$

where $\hat{\psi}(\xi)$ is the Fourier transform of $\psi(x)$. In fact, for all $\xi\in\mathbb{R}^d$, the solution of equation (1) with initial condition $e^{i\boldsymbol{\xi}\cdot\mathbf{x}}$ is of the form

$$T(\mathbf{x},t) = S(\mathbf{x},t,\boldsymbol{\xi})e^{i\boldsymbol{\xi}\cdot\mathbf{x}} \tag{13}$$

where $S(\mathbf{x},t,\mathbf{x})$ is the solution of

$$\left(\frac{\partial}{\partial t} + \mathbf{u}(\mathbf{x})\cdot(\nabla + i\boldsymbol{\xi})\right)S(\mathbf{x},t,\boldsymbol{\xi}) = \sum_{j=1}^{d} D_0\left(\frac{\partial}{\partial x_j} + i\,\xi_j\right)^2 S(\mathbf{x},t,\boldsymbol{\xi}) \tag{14}$$

with initial condition

$$S(\mathbf{x},0,\boldsymbol{\xi}) = 1\quad,\quad \mathbf{x}\in\mathbb{R}^d. \tag{15}$$

From (14), (15) it then follows that $\mathbf{x} \longmapsto S(\mathbf{x}, t, \boldsymbol{\xi})$ is a stationary process. Using this fact and (13) we obtain the representation (11) for all $\psi \in C_0^\infty(\mathbb{R}^d)$, with

$$\overline{P}_z(t, \boldsymbol{\xi}) = < S(\mathbf{0}, t, \boldsymbol{\xi}) > . \tag{16}$$

The function $\overline{G}_z(\lambda, \boldsymbol{\xi})$ in (12) is simply the Laplace transform of $\overline{P}_z(t, \boldsymbol{\xi})$.

We will derive an integral representation for $\overline{G}_z(\lambda, \boldsymbol{\xi})$ which will be useful to study the long-time/large-distance properties of (1). For this we observe that from (1), (9), the operators $\mathbf{G}_z(\lambda)$ and $\mathbf{G}_0(\lambda)$ satisfy the equation

$$\mathbf{G}_z(\lambda) + z\,\mathbf{G}_0(\lambda)\mathbf{u} \cdot \nabla \mathbf{G}_z(\lambda) = \mathbf{G}_0(\lambda). \tag{17}$$

Introducing the operator

$$\mathbf{T}(\lambda) = \mathbf{G}_0(\lambda)\mathbf{u} \cdot \nabla, \tag{18}$$

we obtain, from (17),

$$\mathbf{G}_z(\lambda) = (\mathbf{I} + z\,\mathbf{T}(\lambda))^{-1}\,\mathbf{G}_0(\lambda). \tag{19}$$

We make the crucial observation that $i\mathbf{T}(\lambda)$ is a hermitian operator on $H^1(\mathbb{R}^d; C)$ with the inner product

$$[\psi_1, \psi_2] = \int_{\mathbb{R}^d} \overline{\hat{\psi}_1(\boldsymbol{\xi})}\hat{\psi}_2(\boldsymbol{\xi})(\lambda + D_0|\boldsymbol{\xi}|^2)d\boldsymbol{\xi}, \tag{20}$$

since (using (2))

$$\begin{aligned}[\psi_1, i\,\mathbf{T}(\lambda)\psi_2] &= i\int \overline{\psi_1(\mathbf{x})}\mathbf{u}(\mathbf{x}) \cdot \nabla\psi_2(\mathbf{x})d\mathbf{x} \\ &= -\int \psi_2(\mathbf{x})\mathbf{u}(\mathbf{x}) \cdot \nabla\overline{\psi_1(\mathbf{x})}d\mathbf{x}) \\ &= \overline{[i\,\mathbf{T}(\lambda)\psi_1, \psi_2]}.\end{aligned} \tag{21}$$

[In (20), (21) the overbar denotes complex conjugation]. Hence, from the Spectral Theorem[24], there exists a monotone increasing, right-continuous family of projection operators, $\mathbf{R}(\lambda, \tau)$, such that

$$\mathbf{G}_z(\lambda) = \int_{-\infty}^{+\infty} (1 - iz\tau)^{-1}\,d\mathbf{R}(\lambda, \tau)\mathbf{G}_0(\lambda) \tag{22}$$

for all $z \in C$ such that $Re z \neq 0$. Using (22) we obtain, for all $\psi \in C_0^\infty(\mathbb{R}^d)$,

$$\begin{aligned} &< [\mathbf{G}_0(\lambda)\psi, \mathbf{G}_z(\lambda)\psi] > \\ &= \int_{-\infty}^{+\infty} (1 - iz\tau)^{-1}\, d < [\mathbf{G}_0(\lambda)\psi, \mathbf{R}(\lambda,\tau)\mathbf{G}_0(\lambda)\psi] > \\ &= \int_{-\infty}^{+\infty} (1 - iz\tau)^{-1}\, d\overline{m}(\lambda,\psi,\tau). \end{aligned} \tag{23}$$

The operator $\mathbf{R}(\lambda,\tau)$ being a projection, we conclude that (23) defines a positive measure $d\overline{m}(\lambda,\psi,\cdot)$, such that its successive moments are given by

$$\int_{-\infty}^{+\infty} d\overline{m}(\lambda,\psi,\tau) = (2\pi)^d \int_{\mathbb{R}^d} \frac{|\hat{\psi}(\boldsymbol{\xi})|^2}{\lambda + D_0|\boldsymbol{\xi}|^2} d\boldsymbol{\xi} \tag{24}$$

and

$$\begin{aligned} &\int_{-\infty}^{+\infty} \tau^{2N}\, d\overline{m}(\lambda,\psi,\tau) = (-1)^{N+1}(2\pi)^d \times \\ &\int_{\mathbb{R}^d} \frac{|\hat{\psi}(\boldsymbol{\xi})|^2}{\lambda + D_0|\boldsymbol{\xi}|^2} < \boldsymbol{\xi}\cdot\mathbf{u}(\mathbf{G}_0(\lambda)\mathbf{u}\cdot\nabla)^{2N-2}\,\mathbf{G}_0(\lambda)\mathbf{u}\cdot\boldsymbol{\xi}) > d\boldsymbol{\xi}, \end{aligned} \tag{25}$$

for $N \geq 1$. The odd moments of $d\overline{m}(\lambda,\psi,\cdot)$ vanish because $\mathbf{T}(\lambda)$ is a real, skew-symmetric operator, having a purely imaginary spectrum, symmetric about the origin. Observing that

$$< [\mathbf{G}_0(\lambda)\psi, \overline{\mathbf{G}}_z(\lambda)\psi] >= (2\pi)^d \int_{\mathbb{R}^d} \overline{G}_z(\lambda,\boldsymbol{\xi})|\hat{\psi}(\boldsymbol{\xi})|^2\, d\boldsymbol{\xi}, \tag{26}$$

and using (23), (24), (25), we conclude that for all $\lambda > 0$, $\xi \in \mathbb{R}^d$,

$$\overline{G}_z(\lambda,\boldsymbol{\xi}) = \frac{1}{\lambda + D_0|\boldsymbol{\xi}|^2} \int_{-\infty}^{+\infty} \frac{d\overline{m}(\lambda,\boldsymbol{\xi},\tau)}{1+\tau^2 z^2} \tag{27}$$

where $d\overline{m}(\lambda,\boldsymbol{\xi},\cdot)$ is a probability measure on $\mathbb{R}$, symmetric about $\tau = 0$, such that, for all $N \geq 1$,

$$\begin{aligned} &\int_{-\infty}^{+\infty} \tau^{2N}\, d\overline{m}(\lambda,\boldsymbol{\xi},\tau) \\ &= \frac{(-1)^{N+1}}{\lambda + D_0|\boldsymbol{\xi}|^2} < \boldsymbol{\xi}\cdot\mathbf{u}(\mathbf{G}_0(\lambda)\mathbf{u}\cdot\nabla)^{2N-2}\mathbf{G}_0(\lambda)\mathbf{u}\cdot\boldsymbol{\xi} > . \end{aligned} \tag{28}$$

The representation formula for the Fourier multiplier of the averaged Green's function given in (27) shows that $\overline{G}_z(\lambda,\boldsymbol{\xi})$ is an analytic function of z in the domain $\{z : Rez \neq 0\}$, determined completely by its derivatives at $z = 0$, through the moments of the Stieltjes probability measure $d\overline{m}(\lambda(\xi,\cdot)$ given in (28). This is the main idea behind this perturbative approach to the long-time, large-distance homogenization problem. We can formulate the problem of interest as follows : Given a small parameter $\delta > 0$, find a scaling function $\rho = \rho(\delta)$ such that

$$\lim_{\delta \downarrow 0} (\rho(\delta))^2 \overline{G}_z((\rho(\delta))^2\lambda, \delta\boldsymbol{\xi}) \equiv \overline{G}_z^{eff}(\lambda, \boldsymbol{\xi}) \tag{29}$$

exists and differs both from 0 or λ^{-1}, and compute the limit. Equivalently, we seek to determine the scaling function ρ so that

$$\frac{1}{\lambda + \delta^2\rho^{-2}D_0|\boldsymbol{\xi}|^2} \int_{-\infty}^{+\infty} \frac{d\overline{m}(\rho^2\lambda, \delta\boldsymbol{\xi}, \tau)}{1+\tau^2 z^2} \tag{30}$$

has a non-trivial limit as $\delta \to 0$.

3. Renormalized perturbation theory

Instead of studying the "primitive" expansion (27) and the associated asymptotic limit (29), (30), we perform a convenient transformation. Let $\zeta = z^2$ and set

$$f(\zeta) = \int_{-\infty}^{+\infty} \frac{d\overline{m}(\lambda, \xi, \tau)}{1+\tau^2\zeta}. \tag{31}$$

This function is defined and different from zero on the region $\Omega = C\backslash\{Re\zeta < 0; Im\zeta = 0\}$ and has the property that $Imf(\zeta) < 0$ whenever $Re\zeta > 0$. Defining the new function

$$g(\zeta) = \frac{1}{f(\zeta)} - 1 \quad , \quad \zeta \in \Omega, \tag{32}$$

we observe that $Img(\zeta) > 0$ whenever $Im\zeta > 0$. We conclude from this and a classical result in function theory[25], that there exists a positive, finite measure $d\hat{n}(\lambda, \boldsymbol{\xi}, \cdot)$ such that

$$g(\zeta) = \int_{-\infty}^{+\infty} \frac{\zeta\, d\hat{n}(\lambda, \boldsymbol{\xi}, \tau)}{1+\tau^2\zeta}. \tag{33}$$

Therefore, since $f(\zeta) = (1 + g(\zeta))^{-1}$, if we define $d\hat{n}(\lambda, \boldsymbol{\xi}, \cdot) = (\lambda + D_0|\boldsymbol{\xi}|^2)d\hat{n}(\lambda, \boldsymbol{\xi}, \cdot)$ we can recast (27) in the form

$$\overline{G}_z(\lambda, \boldsymbol{\xi}) = \frac{1}{\lambda + D_0|\boldsymbol{\xi}|^2 + \int_{-\infty}^{+\infty} \frac{z^2 d\overline{n}(\lambda, \boldsymbol{\xi}, \tau)}{1+z^2\tau^2}}. \tag{34}$$

Following Kraichnan[14], we will refer to the Stieltjes integral appearing in (34) as the <u>renormalized representation of</u> $\overline{G}_z(\lambda, \boldsymbol{\xi})$ and $d\overline{n}(\lambda, \boldsymbol{\xi}, \cdot)$ as the <u>renormalized measure</u>.

A more detailed analysis of the expressions appearing on the righ-hand side of (28), for $N \geq 1$, will provide both the motivation for making such transformation as well as explicit formulas for the successive moments of the renormalized measure. For this, we study the expressions

$$A_{2N} \equiv < \xi \cdot \mathbf{u}(\mathbf{G}_0(\lambda)\mathbf{u} \cdot \nabla)^{2N-2}\mathbf{G}_0(\lambda)\mathbf{u} \cdot \xi > \tag{35}$$

using the Fourier decomposition of $\mathbf{u}(\mathbf{x})$ given in (3) and the Fourier multipliers of the operators $\mathbf{G}_0(\lambda)$ and ∇. After some algebraic manipulations, we find that

$$A_{2N} = (-1)^{N-1} \sum_{\mathbf{k}_1+\ldots+k_{2n}=0} \frac{\prod_{j=0}^{2N-1} (\boldsymbol{\xi} + \sum_{\ell=1}^{j} \mathbf{k}_\ell) \cdot \hat{\mathbf{u}}(\mathbf{k}_{j+1})}{\prod_{j=1}^{2N-1} [D_0|\boldsymbol{\xi} + \sum_{\ell=1}^{j} \mathbf{k}_\ell|^2 + \lambda]} \tag{36}$$

(with the convention that $\sum_{\ell=1}^{0} \mathbf{k}_\ell = 0$). Because of the incompressibility condition, which implies $\mathbf{k}_{2N} \cdot \hat{\mathbf{u}}(\mathbf{k}_{2N}) = 0$, and the fact that the sum is taken only over wavevectors such that $\sum \mathbf{k}_\ell = 0$, the factor $(\boldsymbol{\xi} + \sum_{1\leq\ell\leq N-1} \mathbf{k}_\ell) \cdot \hat{\mathbf{u}}(\mathbf{k}_{2N})$ can be replaced by $\boldsymbol{\xi} \cdot \hat{\mathbf{u}}(\mathbf{k}_{2N})$. Hence, only $2N-2$ factors containing the vectors $\mathbf{k}$ appear in the numerator. Despite its apparent complication, expression (36) is mathematically well-defined, because of the regularity assumptions on the field $\mathbf{u}(\mathbf{x})$ and the regularizing properties of the operator $\mathbf{G}_0(\lambda)$. In fact, A_{2N} is the Fourier mode of order zero of the function appearing between brackets in (35).

Let us set

$$\mathbf{S}_j = \sum_{\ell=1}^{j} \mathbf{k}_\ell \quad , \quad 1 \leq j \leq 2N. \tag{37}$$

Among the summands appearing in (36), corresponding to different sets of wavevectors $(\mathbf{k}_1, ...\mathbf{k}_{2N})$ such that $\mathbf{S}_{2N} = 0$, some are such that $\mathbf{S}_{j_0} = 0$ for $j_0 < 2N$. A summand satisfying such property can be factored as follows

$$\frac{\prod_{j=0}^{j_0-1}(\boldsymbol{\xi}+\sum_{\ell=1}^{j}\mathbf{k}_\ell)\cdot\hat{\mathbf{u}}(\mathbf{k}_{j+1})}{\prod_{j=1}^{j_0}[D_0|\boldsymbol{\xi}+\sum_{\ell=1}^{j}\mathbf{k}_\ell|^2+\lambda]}\overline{G}_0(\lambda,\boldsymbol{\xi})\frac{\prod_{j=j_0}^{2N-1}(\boldsymbol{\xi}+\sum_{\ell=j_0+1}^{j}\mathbf{k}_\ell)\cdot\hat{\mathbf{u}}(k_{j+1})}{\prod_{j=j_0+1}^{2N-1}[D_0|\boldsymbol{\xi}+\sum_{\ell=j_0+1}^{j}\mathbf{k}_\ell|^2+\lambda]}. \tag{38}$$

Clearly, both factors in (38) appear as summands of the "lower order" terms A_{j_0} and A_{2N-j_0}. A summand appearing in A_{2N} for which $\mathbf{S}^j = 0$ for $j < 2N$ is termed a reducible diagram in field-theoretic terminology. Summands for which $\mathbf{S}_j \neq 0$ for $j < 2N$ and $\mathbf{S}_{2N} = 0$ are called irreducible diagrams. [We commit here an abuse of language. In field theory, every summand is associated to a (Feynman) diagram or graph, which can be reducible or irreducible. This distinction is irrelevant for our purposes.] Every reducible diagram is a product of irreducible diagrams of lower order and of the Fourier transform of the unperturbed Green function $\overline{G}_0(\lambda, \boldsymbol{\xi}) = (\lambda + D_0|\boldsymbol{\xi}|^2)^{-1}$, as can be checked by repeating the decomposition in (38) to each factor, and continuing this procedure until all subdiagrams that appear are irreducible.

Let B_{2r}, $r = 1, 2, 3...$, denote the sum of all irreducible diagrams of order $2r$, i.e.,

$$B_{2r} = \sum_{\substack{\mathbf{S}_{2r}=0,\, \mathbf{S}_j\neq 0\\ 1\leq j<2r}} \frac{\prod_{j=0}^{2r-1}(\boldsymbol{\xi}+\sum_{\ell=1}^{j}\mathbf{k}_\ell)\cdot\hat{\mathbf{u}}(\mathbf{k}_{j+1})}{\prod_{j=1}^{2r-1}[D_0|\boldsymbol{\xi}+\sum_{\ell=1}^{j}\mathbf{k}_\ell|^2+\lambda]}. \tag{39}$$

Then, using (28), the following formula holds

$$\begin{aligned} &\int \tau^{2N}\, d\overline{m}(\lambda,\boldsymbol{\xi},\tau) = (-1)^{N-1} A_{2N}\, \overline{G}_0(\lambda,\boldsymbol{\xi}) \\ &= \sum_{r_1+...+r_s=N} B_{2r_1}\overline{G}_0(\lambda,\boldsymbol{\xi})B_{2r_2}...\overline{G}_0(\lambda,\boldsymbol{\xi})B_{2r_s}\overline{G}_0(\lambda,\boldsymbol{\xi}), \end{aligned} \tag{40}$$

which expresses the moments of $d\overline{m}$ only in terms of sums of irreducible diagrams.

Let us expand the Stieltjes integral in (34), formally in powers of z^2 :

$$\int_{-\infty}^{+\infty} \frac{z^2\, d\overline{n}(\lambda,\boldsymbol{\xi},\tau)}{1+\tau^2 z^2} \cong \sum_{r=1}^{\infty} z^{2r}\, C_{2r} \tag{41}$$

where $C_{2r} = (-1)^r \int \tau^{2(r-1)} d\overline{n}$. Substituting this series in (34) and expanding, we obtain

$$\frac{\overline{G}_z(\lambda,\boldsymbol{\xi})}{\overline{G}_0(\lambda,\boldsymbol{\xi})} = 1 + \sum_{N=1}^{\infty} (-1)^N z^{2N} \sum_{r_1+\ldots+r_s=N} C_{2r_1}\,\overline{G}_0(\lambda,\boldsymbol{\xi})\ldots C_{2r_s}\,\overline{G}_0(\lambda,\boldsymbol{\xi}). \quad (42)$$

This can be compared with the expansion obtained directly from (27), yielding

$$A_{2N} = -\sum_{r_1+\ldots+r_s=N} C_{2r_1}\,\overline{G}_0(\lambda,\xi)\ldots C_{2r_2}\ldots\overline{G}_0(\lambda,\xi) C_{2r_s}\ ,\ N \geq 1. \quad (43)$$

From this formula and (40) we conclude finally that

$$C_{2r} = (-1)^r\, B_{2r} \quad (44)$$

and hence that

$$\int_{-\infty}^{+\infty} \tau^{2(r-1)} d\overline{n}(\lambda,\boldsymbol{\xi},\tau) = B_{2r} \quad , \quad r \geq 1. \quad (45)$$

In summary, we have shown that for each $r \geq 1$, the moment of order $r-1$, of the renormalized Stieltjes measure $d\overline{n}(\lambda,\xi,\cdot)$ is given by B_{2r}, the sum of all irreducible diagrams of order $2r$.

4. Homogenization and mean-field régime

In this section we show that if the velocity field $\mathbf{u}(\mathbf{x})$ satisfies the mean-field condition

$$\sum_{\mathbf{k}} \frac{< |\hat{\mathbf{u}}(\mathbf{k})|^2 >}{|\mathbf{k}|^2} < +\infty \quad (46)$$

and an additional technical assumption, then

$$\lim_{\delta \downarrow 0} \delta^2\, \overline{G}_z(\delta^2\lambda, \delta\boldsymbol{\xi}) = \frac{1}{\lambda + \boldsymbol{\xi}^T \cdot \mathbf{D}^* \cdot \boldsymbol{\xi}}, \quad (47)$$

where $\mathbf{D}^*$ is the homogenized, or effective diffusivity tensor. This result is a weak version of the homogenization theorem for the initial-value problem associated with (1), in the long-time, large-distance limit : the solution $T_\delta(\mathbf{x},t)$ of the problem

$$\begin{cases} \dfrac{\partial}{\partial t} T_\delta(\mathbf{x},t) + \dfrac{1}{\delta}\,\mathbf{u}(\dfrac{x}{\delta}) \cdot \nabla T_\delta(\mathbf{x},t) = D_0\, \Delta T_\delta(\mathbf{x},t)\ ,\ \mathbf{x} \in \mathbb{R}^d\ ,\ t > 0 \\ T_\delta(x, t=0) = T_0(x) \quad , \quad x \in \mathbb{R}^d \end{cases} \quad (48)$$

satisfies

$$< T_\delta(\mathbf{x},t) > \to \overline{T}(x,t) \tag{49}$$

weakly in $L^2_{loc}(\mathbb{R}^d \times \mathbb{R}^+)$, where $\overline{T}(x,t)$ is the solution of

$$\begin{cases} \frac{\partial}{\partial t}\overline{T}(x,t) = \sum_{\ell\ell'} D^*_{\ell\ell'} \frac{\partial^2}{\partial x_\ell \, \partial x_{\ell'}} \overline{T}(\mathbf{x},t) \\ \overline{T}(x,t=0) = T_0(x). \end{cases} \tag{50}$$

Probabilistic and P.D.E. proofs of this theorem have been given before by several authors[22,23,28,11].

To prove (47) in the present framework, we define the skew-symmetric matrix-valued field $\mathbf{H}(\mathbf{x})$ by

$$\mathbf{H}(\mathbf{x}) = \sum_{\mathbf{k}} \hat{\mathbf{H}}(\mathbf{k}) e^{i\mathbf{k}\cdot\mathbf{x}} \tag{51}$$

with

$$\hat{H}_{\alpha\beta}(\mathbf{k}) = \frac{1}{i|\mathbf{k}|^2}[k_\alpha \, \hat{u}_\beta(\mathbf{k}) - k_\beta \, \hat{\mathbf{u}}_\alpha(\mathbf{k})] \tag{51'}$$

for $1 \le \alpha,\ \beta \le d$. Of course, the summation in (51) extends over the (discrete) support of $\mathbf{u}(\mathbf{x})$ defined in (3). Note that $\mathbf{H}(\mathbf{x})$ is a well-defined field since :

$$\sum_{\mathbf{k}} < |\hat{\mathbf{H}}(\mathbf{k})|^2 > = 2\sum_{\mathbf{k}} \frac{< |\hat{\mathbf{u}}(\mathbf{k})|^2 >}{|\mathbf{k}|^2} < +\infty. \tag{52}$$

From (51'), we see that

$$\nabla \cdot \mathbf{H}(\mathbf{x}) = \mathbf{u}(\mathbf{x}). \tag{53}$$

In addition to the mean-field condition (46), which guarantees that $\mathbf{H} \in L^2(< \cdot >)$, we will assume that

$$< |\mathbf{H}(\mathbf{x})|^p > < +\infty, \quad \text{for all } p \ge 1, \tag{54}$$

and that for all $p > 1$, $\varepsilon > 0$, there exists a finite set F of wave-vectors $\mathbf{k}$ such that

$$< |\mathbf{H}(\mathbf{x}) - \sum_{\mathbf{k}\in F} \hat{\mathbf{H}}(\mathbf{k})\, e^{i\mathbf{k}\cdot\mathbf{x}}|^p > \le \varepsilon. \tag{54'}$$

Using the field $\mathbf{H}(\mathbf{x})$ and equation (53), we can rewrite the moments of the primitive expansion, as

$$\int_{-\infty}^{+\infty} \tau^{2N}\, d\overline{m}(\lambda, \boldsymbol{\xi}, \tau) = \frac{(-1)^{N+1}}{\lambda + D_0|\boldsymbol{\xi}|^2}\, \boldsymbol{\xi}^T \cdot < \mathbf{H}(\boldsymbol{\Gamma}(\lambda)\mathbf{H})^{2N-1} > \cdot \boldsymbol{\xi} \qquad (55)$$

where the operator $\boldsymbol{\Gamma}(\lambda)$ is defined by

$$\Gamma_{\alpha\beta}(\lambda) = (\lambda - D_0\Delta)^{-1}\, \frac{\partial^2}{\partial x_\alpha\, \partial x_\beta}. \qquad (56)$$

Making the change of variables $\boldsymbol{\xi} \rightarrow \delta\boldsymbol{\xi}$ and $\lambda \rightarrow \delta^2\lambda$ we obtain, from (55),

$$\begin{aligned} &\int_{-\infty}^{+\infty} \tau^{2N}\, d\overline{m}(\delta^2\lambda, \delta\boldsymbol{\xi}, \tau) \\ &= \frac{(-1)^{N+1}}{\lambda + D_0|\boldsymbol{\xi}|^2}\, \boldsymbol{\xi}^T \cdot < \mathbf{H}(\boldsymbol{\Gamma}(\delta^2\lambda)\mathbf{H})^{2N-1} > \cdot \boldsymbol{\xi}. \end{aligned} \qquad (57)$$

For instance, for $N = 1$, we obtain

$$\begin{aligned} &\int_{-\infty}^{+\infty} \tau^2\, d\overline{m}(\delta^2\lambda, \delta\boldsymbol{\xi}, \tau) \\ &\quad = \frac{1}{\lambda + D_0|\boldsymbol{\xi}|^2}\, \xi^T \cdot < \mathbf{H}\, \boldsymbol{\Gamma}(\delta^2\lambda)\mathbf{H} > \xi \\ &\quad = \frac{1}{\lambda + D_0|\boldsymbol{\xi}|^2} \sum_{\mathbf{k}} \frac{|\xi \cdot \hat{\mathbf{u}}(\mathbf{k})|^2}{\delta^2\lambda + |\delta\boldsymbol{\xi} + \mathbf{k}|^2}, \end{aligned} \qquad (58)$$

which is bounded uniformly in δ if and only if the mean-field condition (46) holds. We claim that the right-hand side of (57) is bounded independently of δ, for all $N \geq 1$. This is a consequence of the assumption (54) on the higher integrability of $\mathbf{H}(\mathbf{x})$ and of the fact that the operator $\boldsymbol{\Gamma}(\delta^2\lambda)$ corresponds to the Fourier multiplier

$$\hat{\Gamma}_{\alpha\beta}(\delta^2\lambda; \mathbf{k}) = \frac{-k_\alpha\, k_\beta}{\delta^2\lambda + D_0|\mathbf{k}|^2}, \qquad (59)$$

which defines, (as a consequence of the classical Mihlin L^p multiplier theorem[26]) a bounded operator from $L^p(< \cdot >)$ to $L^p(< \cdot >)$ for all $p > 1$. For each p, the operator norm is bounded independently of δ.

Using the uniform boundedness of the moments of the measure $d\overline{m}(\delta^2\lambda, \delta\boldsymbol{\xi}, \cdot)$ and (54'), it is possible to reduce the problem to the case of a field $\mathbf{H}(\mathbf{x})$ which has a Fourier transform $\{\hat{\mathbf{H}}(\mathbf{k})\}$, supported on a finite set of wavevectors $\mathbf{k}$, $|\mathbf{k}| \neq 0$. For the sake of brevity we omit the details of the argument leading to this reduction, based on approximating the Stieltjes integral (27) by Padé (rational function) approximants.

Finally, we compute the limit of $\delta^2 \overline{G}_z(\delta^2\lambda, \delta\boldsymbol{\xi})$ as $\delta \to 0$, assuming that $\mathbf{H}(\mathbf{x})$ has finitely supported Fourier transform. For this, we use the renormalized representation (34). As show in section 2, the moments of the renormalized measure $d\overline{n}(\delta^2\lambda, \delta\boldsymbol{\xi}, \cdot)$ correspond to sums of irreducible diagrams. Using the equation (53), we can express such diagrams in terms of the $\mathbf{H}$-field.

In fact, we have, from (39), (53),

$$B_{2r} = (-1)^r \sum_{\substack{\mathbf{S}_{2r}=0, \mathbf{S}_j \neq 0 \\ 1 \leq j > 2r}} \boldsymbol{\xi}^T \cdot \hat{\mathbf{H}}(\mathbf{k}_1) \left[\prod_{j=1}^{2r-1} \hat{\Gamma}(\lambda, \boldsymbol{\xi} + \mathbf{S}_j) \hat{\mathbf{H}}(\mathbf{k}_{j+1}) \right] \cdot \boldsymbol{\xi}. \tag{60}$$

Making the change of variables $\lambda \Longleftrightarrow \delta^2\lambda, \boldsymbol{\xi} \Longleftrightarrow \delta\boldsymbol{\xi}$ and using the results of chapter 3, we obtain

$$\begin{aligned} &\frac{1}{\delta^2} \int_{-\infty}^{+\infty} \tau^{2(r-1)} d\overline{n}(\delta^2\lambda, \delta\boldsymbol{\xi}, \tau) = \\ &= \sum_{\substack{\mathbf{S}_{2r}=0, \mathbf{S}_j \neq 0 \\ 1 \leq j < 2r}} \boldsymbol{\xi}^T \cdot \hat{\mathbf{H}}(\mathbf{k}_1) \left[\prod_{j=1}^{2r-1} \hat{\Gamma}(\delta^2\lambda, \delta\boldsymbol{\xi} + \mathbf{S}_j) \hat{\mathbf{H}}(\mathbf{k}_{j+1}) \right] \cdot \boldsymbol{\xi}. \end{aligned} \tag{61}$$

This formula shows that for all $r \geq 1$ the rth-moments of $\delta^{-2} d\overline{n}(S^2\lambda, \delta\boldsymbol{\xi}, \cdot)$ are bounded independently of δ, i.e.

$$\frac{1}{\delta^2} \int \tau^{2r} d\overline{n}(\delta\lambda, \delta\boldsymbol{\xi}, \tau) \leq C_r < \infty. \tag{62}$$

In particular, $\{\delta^{-2} d\overline{n}(\delta^2\lambda, \delta\boldsymbol{\xi}, \cdot)\}$ is relatively compact in the topology of weak convergence of measures (tight). Finally, note that, since $S_j \neq 0$ for $1 \leq j < 2r$ in (61),

$$\begin{aligned} \lim_{\delta \downarrow 0} \hat{\Gamma}(\delta^2\lambda, \delta\boldsymbol{\xi} + \mathbf{S}_j) &= \hat{\Gamma}(0, \mathbf{S}_j) \\ &= -\frac{\mathbf{S}_j \mathbf{S}_j^T}{|S_j|^2}. \end{aligned} \tag{63}$$

Note also that $\hat{\Gamma}(0,\mathbf{k})$ can be interpreted as the Fourier multiplier of the operator $\boldsymbol{\Gamma} = (-\Delta)^{-1}\nabla\nabla$, defined on stationary fields having mean zero. We conclude from (61), (62), (63), that

$$\begin{aligned}\lim_{\delta\downarrow 0} &\delta^2 \overline{G}'_z(\delta^2\lambda, \delta\boldsymbol{\xi}) \\ &= \lim_{\delta\downarrow 0}\left[\lambda + D_0|\boldsymbol{\xi}|^2 + \frac{1}{\delta^2}\int_{-\infty}^{+\infty}\frac{z^2\, d\overline{n}(\delta^2\lambda, \delta\boldsymbol{\xi}, \tau)}{1+z^2\tau^2}\right]^{-1} \\ &= \frac{1}{\lambda + \boldsymbol{\xi}^T\cdot D^*\boldsymbol{\xi}},\end{aligned} \tag{64}$$

where the effective diffusivity tensor $\mathbf{D}^*$ is given by the formula

$$\mathbf{D}^* = D_0\left[\mathbf{I} + \int_{-\infty}^{+\infty}\frac{(z/D_0)^2 d\boldsymbol{\nu}(\tau)}{1+\tau^2(z/D_0)^2}\right], \tag{65}$$

in which $d\boldsymbol{\nu}(\tau)$ is a positive-definite tensor-valued measure determined completely by the statistics of $\mathbf{H}(\mathbf{x})$ (or, equivalently, $\mathbf{u}(\mathbf{x})$). The measure $\boldsymbol{\xi}^T\cdot d\boldsymbol{\nu}(\cdot)\cdot\boldsymbol{\xi}$ is, in fact, the weak limit of the family $\delta^{-2}d\overline{n}(\delta^2\lambda, \delta\boldsymbol{\xi}, \cdot)$ as $\delta \to 0$. The moments of this measure are, from (62), (63), given by

$$\int_{-\infty}^{+\infty}\tau^{2(N-1)}d\boldsymbol{\nu}(\tau) = (-1)^{N+1} < \mathbf{H}(\boldsymbol{\Gamma}\mathbf{H})^{2N-1} > \tag{66}$$

for $N \geq 1$.

This concludes the proof of the homogenization result. The Stieltjes integral representation for the effective diffusivity arising in mean field transport was first derived in Refs. 10, 11.

5. Infrared divergence and anomalous scaling laws

In this section we present an application of the perturbation method to the calculation of the long-time/large distance scaling laws and effective equations that arise when the mean-field condition (46) is violated, i.e.,

$$\sum_{\mathbf{k}}\frac{< |\hat{\mathbf{u}}(\mathbf{k})|^2 >}{|\mathbf{k}|^2} = +\infty. \tag{67}$$

In such cases, the velocity has long-range correlations that influence the dynamics at long times. The crossover that takes place can be explained using a Lagrangian argument. In fact, a particle undergoing a continuous-time random walk with transition probabilities $P_z(t, x, y)$ with not "feel" the advective term at large distances if (46) holds. On the other hand, long-range correlations that arise with (67) can cause the particle to move under the influence of the advective field over long distances, so that the typical displacement at long times grows much faster than the square-root of the elapsed time. In terms of the scaling function, this leads to $\rho(\delta) \ll \delta$.

The analysis of the anomalous régime depends in a sensitive way on the behavior of the Fourier transform of $\mathbf{u}(\mathbf{x})$ near $|\mathbf{k}| \ll 1$, the infrared region. In particular, the approximation of the random field by one with a finite support in k-space (done via the $\mathbf{H}$-field in section 4) can no longer be done, as one expects that the dynamics to be dominated precisely by the infrared region, in an arbitrarily small neighborhood of the origin.

To present our method in this context we analyze a simple shear flow with a quasiperiodic coefficient,

$$\mathbf{u}(\mathbf{x}) = \begin{bmatrix} 0 \\ u(x_1) \end{bmatrix} \qquad \mathbf{x} = (x_1, x_2) \tag{68}$$

where

$$u(x_1) = \sum_{n=1}^{+\infty} \theta^{n\beta} \cos(\theta^n x_1), \tag{69}$$

θ is a real parameter with $0 < \theta < \frac{1}{3}$ and $\beta \in (0, 1]$. The parameter β regulates the strength of the Fourier transform in the infrared. Note that

$$\hat{u}(\mathbf{k}) = \frac{1}{2}\theta^{n\beta} \quad , \quad \mathbf{k} = \pm\theta^n \begin{pmatrix} 1 \\ 0 \end{pmatrix} = \pm\theta^n \mathbf{e}_1 \tag{70}$$

so that, for $|\mathbf{k}| = \theta^n$,

$$\frac{|\hat{\mathbf{u}}(\mathbf{k})|^2}{|\mathbf{k}|^2} = \theta^{2n(\beta-1)} \tag{71}$$

Thus, for $\beta > 1$, the mean-field condition is satisfied, while $0 < \beta \leq 1$ corresponds to (67).

We wish to determine the scaling parameter $\rho = \rho(\delta)$ so that

$$\lim_{\delta \downarrow 0} \frac{1}{\rho^2} \int_{-\infty}^{+\infty} d\overline{n}(\rho^2\lambda, \delta\xi\mathbf{e}_2; \tau) \tag{72}$$

is finite and positive, with $\mathbf{e}_2 = \begin{pmatrix} 0 \\ 1 \end{pmatrix}$. This particular choice of wave-vector $\boldsymbol{\xi} = \xi \mathbf{e}_2$ is made because of the anisotropy of $\mathbf{u}(\mathbf{x})$. A straightforward calculation yields, from (45),

$$\begin{aligned} &\frac{1}{\rho^2} \int_{-\infty}^{+\infty} d\overline{n}(\rho^2 \lambda, \delta\xi \mathbf{e}_2, \tau) \\ &= \frac{\delta^2}{4\rho^2} \sum_{n=1}^{+\infty} \frac{\theta^{2n\beta} \xi^2}{D_0 \delta^2 \xi^2 + D_0 \theta^{2n} + \rho^2 \lambda} \\ &= \frac{\delta^2}{4\rho^4} \sum_{n=1}^{+\infty} \frac{\theta^{2n\beta} \xi^2}{D_0 \delta^2 \rho^{-2} \xi^2 + D_0 \rho^{-2} \theta^{2n} + \lambda}. \end{aligned} \tag{73}$$

Setting $\rho = \theta^M$, where M is a large positive integer and anticipating that $\rho\delta^{-1} \to 0$ as $\delta \to 0$, we obtain

$$\begin{aligned} &\frac{1}{\rho^2} \int_{-\infty}^{+\infty} d\overline{n}(\rho^2 \lambda, \delta\xi \mathbf{e}_2, \tau) \\ &= \frac{\delta^2 \theta^{2M\beta}}{4\rho^4} \sum_{n=-M+1}^{+\infty} \frac{\theta^{2n\beta} \xi^2}{D_0 \delta^2 \rho^{-2} \xi^2 + D_0 \theta^{2n} + \lambda} \cdot \\ &\sim \frac{\delta^2}{4\rho^{4-2\beta}} \sum_{n=-M+1}^{+\infty} \frac{\theta^{2n\beta} \xi^2}{D_0 \theta^{2n} + \lambda} \end{aligned} \tag{74}$$

We distinguish two cases :

A) $\underline{0 < \beta < 1}$. Then the series in (74) is uniformly convergent as $M \to +\infty$. Hence, making the choice

$$\rho(\delta) = \delta^{\frac{1}{2-\beta}} \tag{75}$$

we obtain

$$\begin{aligned} &\lim_{\delta \downarrow 0} \frac{1}{\rho^2} \int_{-\infty}^{+\infty} d\overline{n}(\rho^2 \lambda, \delta\xi \mathbf{e}_2, \tau) \\ &= \frac{1}{4} \sum_{n=-\infty}^{+\infty} \frac{\theta^{2n\beta} \xi^2}{D_0 \theta^{2n} + \lambda}. \end{aligned} \tag{76}$$

B) $\underline{\beta = 1}$. We have

$$\sum_{-M+1}^{+\infty} \frac{\theta^{2n} \xi^2}{D_0 \theta^{2n} + \lambda} \sim \frac{M}{D_0}. \tag{77}$$

Therefore, setting

$$\rho(\delta) = \delta(\log \frac{1}{\delta})^{1/2} \tag{78}$$

and recalling that $M = \frac{\log \rho}{\log \theta}$, we obtain

$$\lim_{\delta \downarrow 0} \frac{1}{\rho^2} \int_{-\infty}^{+\infty} d\overline{n}(\rho^2\lambda, \delta\xi\mathbf{e}_2, \tau) = \frac{\xi^2}{4D_0|\log \theta|}. \tag{79}$$

We apply these scaling results to compute the limit of $\rho^2 G_z(\rho^2\lambda, \delta\boldsymbol{\xi})$ as $\delta \to 0$, using the renormalized expansion (34). Accordingly

$$\begin{aligned} &\lim_{\delta \downarrow 0} \rho^2 \overline{G}_z(\rho^2\lambda, \delta\xi\mathbf{e}_2) \equiv \overline{G}_z^{eff}(\lambda, \xi) \\ &= \lim_{\delta \downarrow 0} \left[\lambda + \delta^2\rho^{-2}\xi^2 + \rho^{-2} \int_{-\infty}^{+\infty} \frac{z^2 d\overline{n}(\rho^2\lambda, \delta\xi\mathbf{e}_2, \tau)}{1 + z^2\tau^2}\right]^{-1} \\ &= \left[\lambda + \int_{-\infty}^{+\infty} \frac{z^2 d\overline{\nu}(\lambda, \xi, \tau)}{1 + z^2\tau^2}\right]^{-1} \end{aligned} \tag{80}$$

where, $d\overline{\nu}(\lambda, \xi, \cdot)$ is the limit in the sense of the vague topology of measures of the family $\frac{1}{\rho^2} d\overline{n}(\rho^2\lambda, \delta\xi\mathbf{e}_2, \cdot)$ as $\delta \to 0$. To verify that the scaling laws determined in (75), (78) are correct, one must check that the measures $d\overline{\nu}$ are not equal to zero (in which case we would obtain the trivial limit $\lim_{\delta \downarrow 0} \rho^2 \overline{G}_z(\rho\lambda, \delta\xi\mathbf{e}_2) = 1/\lambda!$). This crucial step is done by checking that the second moments of $\rho^{-2} d\overline{n}(\rho^2\lambda, \delta\xi\mathbf{e}_2, \cdot)$ are uniformly bounded in δ, and hence that the sequence of measures is tight.

As before, we obtain from (45)

$$\begin{aligned} &\frac{1}{\rho^2} \int_{-\infty}^{+\infty} \tau^2 d\overline{n}(\rho^2\lambda, \delta\xi\mathbf{e}_2, \tau) \\ &= \frac{1}{\rho^2} \sum_{\substack{\mathbf{S}_4=0 \\ \mathbf{S}_j \neq 0, j<4}} \frac{\prod_{j=0}^{3} (\delta\xi\mathbf{e}_2 + \mathbf{S}_j)\hat{u}(\mathbf{k}_{j+1})}{\prod_{j=1}^{3} [D_0\delta^2\xi^2 + C_0|\mathbf{S}_j|^2 + \rho^2\lambda]} \\ &\cong \frac{\xi^4\delta^4}{16\rho^{4(2-\beta)}} \sum_{\substack{\mathbf{S}_4=0, \mathbf{S}_j \neq 0, j<4 \\ |\mathbf{k}_j| \leq \theta^M}} \frac{\prod_{j=1}^{4} |\mathbf{k}_j|^\beta}{\prod_{j=1}^{3} [D_0|\mathbf{S}_j|^2 + \lambda]} \end{aligned} \tag{81}$$

To obtain this formula, we used the special form of the Fourier transform of $u(x_1)$, through equation (70), and neglected the terms $D_0\delta^2\rho^{-2}\xi^2$ in the denominators, since $\delta^2\rho^{-2} \to 0$. We wish to analyze more precisely the diagrams of fourth order in (81). Note that the vectors $\mathbf{k}_j = \pm\theta^{m_j}\,\mathbf{e}_1$, $1 \leq j \leq 4$, with $m_j \geq -M$, satisfy

$$\mathbf{k}_1 + \mathbf{k}_2 + \mathbf{k}_3 + \mathbf{k}_4 = 0. \tag{82}$$

We claim that this sum can be equal to zero only if

$$\begin{gathered} \mathbf{k}_1 + \mathbf{k}_3 = 0 \quad ; \quad \mathbf{k}_2 + \mathbf{k}_4 = 0 \\ \text{or} \\ \mathbf{k}_1 + \mathbf{k}_4 = 0 \quad ; \quad \mathbf{k}_2 + \mathbf{k}_3 = 0. \end{gathered} \tag{83}$$

Indeed, the case

$$\mathbf{k}_1 + \mathbf{k}_2 = 0 \quad ; \quad \mathbf{k}_3 + \mathbf{k}_4 = 0 \tag{84}$$

is ruled out because the diagram must be irreducible. Moreover, if (82) holds then

$$\theta^{m_1} \pm \theta^{m_2} \pm \theta^{m_3} \pm \theta^{m_4} = 0. \tag{85}$$

Suppose that $m_1 < m_j$ for $j > 1$. Then (85) implies

$$1 + \theta^{n_1} \pm \theta^{n_2} \pm \theta^{n_3} = 0 \tag{86}$$

with $n_i \geq 1$. Consequently,

$$1 < 3\theta^{\min(n_1,n_2,n_3)} \tag{87}$$

contradicting the assumption $\theta < 1/3$. We conclude that $m_1 = m_j$ for $j \neq 1$, and hence $|\mathbf{k}_1| = |\mathbf{k}_j|$. Arguing further in this way, it is easy to check that one of the two possibilities of (83) must indeed occur (statistically, this means that the four-point correlation function of $u(x_1)$ splits, as in the Gaussian case, into a sum of products of two-point correlation functions). We can write a typical summand ocurring in the last line of (81) as either

$$\frac{1}{16}\,\frac{k_1^2\,k_2^2}{(D_0k_1^2 + \lambda)[D_0(k_1 + k_2)^2 + \lambda](D_0k_2^2 + \lambda)} \tag{88}$$

which corresponds to $k_1+k_3=0$, $k_2+k_4=0$, or

$$\frac{1}{16}\frac{k_1^2k_2^2}{(D_0k_1^2+\lambda)(D_0(k_1+k_2)^2+\lambda)(D_0k_1^2+\lambda)} \tag{89}$$

corresponding to $k_1+k_4=0$, $k_2+k_3=0$ respectively. Clearly, the first expression is bounded by

$$\frac{1}{16\lambda}\frac{k_1^2}{(D_0k_1^2+\lambda)}\cdot\frac{k_2^2}{(D_0k_2^2+\lambda)}. \tag{90}$$

The expression (89) can be written in the form

$$\frac{1}{16\lambda}\cdot\frac{k_1^2}{(D_0k_1^2+\lambda)}\cdot\frac{k_2^2}{(D_0k_2^2+\lambda)}\cdot\frac{D_0k_2^2+\lambda}{(D_0k_1^2+\lambda)(D_0(k_1+k_2)^2+\lambda)}. \tag{91}$$

The factor appearing on the right of this quantity is such that

$$\begin{aligned}&\frac{D_0k_2^2+\lambda}{(D_0k_1^2+\lambda)(D_0(k_1+k_2)^2+\lambda)}\\ &\quad\le\frac{2[(D_0k_1^2+\lambda)+D_0(k_1+k_2)^2+\lambda]}{(D_0k_1^2+\lambda)(D_0(k_1+k_2)^2+\lambda)}\le\frac{4}{\lambda},\end{aligned} \tag{92}$$

and hence, from (81), the second moment of $\rho^{-2}d\overline{n}(\rho^2\lambda,\delta\xi\mathbf{e}_2,\cdot)$ satisfies

$$\begin{aligned}&\frac{1}{\rho^2}\int\tau^2\,d\overline{n}(\rho^2\lambda,\delta\xi\mathbf{e}_2;\tau)\\ &\quad\le const.\frac{\delta^4\xi^4}{\rho^{4(2-\beta)}}\times\left(\sum_{n=-M+1}^{+\infty}\frac{\theta^{2n\beta}}{D_0\theta^{2n}+\lambda}\right)^2,\end{aligned} \tag{93}$$

where $M\cong|\log\rho|/|\log\theta|$. This shows that the second moments are uniformly bounded as $\delta\to0$ in both cases $\beta\in(0,1)$ and $\beta=1$. We conclude that the limiting measures appearing in the formula (80) are non trivial, with total mass given by the right-hand side of (76) or (79), according to whether $\beta\in(0,1)$ or $\beta=1$.

Finally, we determine more precisely the averaged Green function using (80) and dimensional arguments. Indeed, from (80),

$$\overline{G}_z^{eff}(\lambda,\xi)=\lambda^{-1}\overline{G}_z^{eff}(1,\xi\lambda^{-\frac{2-\beta}{2}}) \tag{94}$$

for $\beta\in(0,1]$. Also, from the form of the basic equation (1),

$$\overline{G}_z^{eff}(\lambda,\xi)=\frac{1}{D_0}\overline{G}_{\frac{z}{D_0},1}^{eff}(\lambda/D_0,\xi) \tag{95}$$

where $\overline{G}^{eff}_{z,1}(\lambda, \xi)$ is the effective averaged Green function for (1) with $D_0 = 1$. Equations (94), (95) imply, using (80), that $\overline{G}^{eff}_{z}(\lambda, \xi)$ is of the form

$$\overline{G}^{eff}_{z}(\lambda, \xi) = \frac{1}{\lambda}\left[1 + \int_{-\infty}^{+\infty} \frac{(z/D_0)^2\, d\overline{\nu}(\xi\lambda^{-\frac{2-\beta}{2}} D_0^{\frac{2-\beta}{2}}; \tau)}{1 + (z/D_0)^2\, \tau^2}\right]^{-1}, \tag{96}$$

where $d\overline{\nu}(\alpha;\tau)$ is, for each α, a positive measure, symmetric about $\tau = 0$.

This expression for the averaged Green function can be simplified even further. In fact, it is possible to show that $\overline{G}^{eff}_{z}(\lambda, \xi)$ remains unchanged under the transformation

$$\xi = \frac{1}{r}\xi' \quad , \quad z = r^{1+\beta} z' \quad , \quad D_0 = r^2 D_0', \tag{97}$$

for any real $r > 0$. As the proof of this fact is somewhat lengthly, it will be given elsewhere[27]. It uses the precise form of the Fourier transform of $u(x_1)$ and the fact that $\mathbf{u}$ is a simple shear flow. The heuristic reason for the invariance of $\overline{G}^{eff}_{z}(\lambda, \xi)$ under the scaling transformation (97) is that the effective equation in the long-time, large distance limit is independent of the modes $\hat{u}(k)$ with $|k| \geq r_0$ for any positive r_0, and depends only on the Fourier transform of $u(x_1)$ in an infinitesimal neighborhood of $k = 0$.

Using this last scaling transformation in formula (96), we obtain

$$\overline{G}^{eff}_{z}(\lambda, \xi) = \frac{1}{\lambda}\left[1 + \int_{-\infty}^{+\infty} \frac{\xi^2\, \lambda^{-2+\beta}\, D_0^{-\beta}\, z^2\, d\overline{\nu}(\tau)}{1 + \xi^2\, \lambda^{-2+\beta}\, D_0^{-\beta}\, z^2\, \tau^2}\right]^{-1} \tag{98}$$

where $d\overline{\nu}(\tau)$ is a positive, finite symmetric measure on the real line. Thus we have $\overline{G}^{eff}_{z}(\lambda, \xi) = [\lambda + D^*(\lambda, \xi)\xi^2]^{-1}$, where the quantity

$$D^*(\lambda, \xi) = \int_{-\infty}^{+\infty} \frac{\lambda^{-1+\beta}\, D_0^{-\beta}\, z^2\, d\overline{\nu}(\tau)}{1 + \xi^2\, \lambda^{-2+\beta}\, D_0^{-\beta}\, z^2\, \tau^2} \tag{99}$$

can be interpreted as a frequency/wave number-dependent diffusivity associated with the long-time large distance limit in the anomalous régime. The reader is referred to refs. 21 and 27 for details.

We conclude by stressing that the procedure used here for the computation of $\overline{G}^{eff}_{z}(\lambda, \xi)$ applies to a variety of anamalous simple shear flows, including the time-independent Gaussian shear flows which are discussed in Reference 21.

Acknowledgements. This work was supported through Grants ARO-DAAL03-89-KJ-0039, AFSOR-90-0090 and NSF-DMS-8802739 (Marco Avellaneda) and ARO-DAAL03-89-K-0013, ONR N00014-89-J-1044 and NSF-DMS-8702864 (Andrew J. Majda). Marco Avellaneda gratefully acknowledges the hospitality of the Laboratoire d'Analyse Numérique, Université Paris VI, during the Spring semester 1990.

References

1. M. Lax, *Rev. Mod. Phys. 23*, 287 (1951).
2. M. Bixon and R. Zwanzig, *J. Chem. Phys. 75*, 2354 (1981).
3. T.R. Kirkpatrick, *J. Chem. Phys. 76*, 4255 (1982).
4. G.H. Fredrickson and E.S.G. Shaqfeh, *Phys. Fluids A* Vol.1, 1, January 1989.
5. M. Beran, *Statistical Continuum Theories*, Wiley-Interscience, New York (1968).
6. J.R. Willis, in *Advances in Applied Mechanics, 21*, p. 1-78, C-S. Yih, Ed. Academic Press, New York, 1981.
7. D.G. Bergman, *Phys. Rep. Phys. Lett. C., 43*, 377 (1978).
8. G.W. Milton, *J. Appl. Phys. 42*, 5286 (1982).
9. K. Golden and G.C. Papanicolaou, *Comm. Math. Phys. 90*, 470 (1983).
10. M. Avellaneda and A.J. Majda, *Phys. Rev. Lett. 62*, 753 (1989).
11. M. Avellaneda and A.J. Majda, preprint, submitted to Physica D, January 1990.
12. P.G. Wolynes, *Phys. Rev. A 11*, 1700 (1975).
13. see articles in *The Padé Approximant in Theoretical Physics*, G.A. Baker, Jr., and J.L. Gammel, eds., Academic Press, New York, 1970.
14. R. Kraichnan, in Ref. 13, p. 129.
15. L. Tartar, in *Essays of Mathematical Analysis in Honour of E. de Giorgi*, Birkhauser, 1989.
16. Y. Amirat, K. Hamdache and A. Ziani, *Ann. Inst. Henri Poincaré*, Anal. non linéaire *6* (5), 397 (1989).
17. Y. Amirat, K. Hamdache and A. Ziani, preprint, January 1990.
18. G. Mathéron and G. de Marsily, *Water Resources Research, 16* (5), 901 (1980).
19. J.P. Bouchaud, A. Comptet, A. Georges, P. Le Doussal, *Journ. Phys.* (Paris) *48*, 1445 (1987).
20. D.L. Koch and J.F. Brady, *Phys. Fluids A, 1*, 47 (1989).

21. M. Avellaneda and A. Majda, preprint, Dec. 1989, to appear in Comm. Math. Phys.
22. K. Oelschläger, *Annals of Prob. 16* (3), 1084 (1988).
23. G.C. Papanicolaou and S.R.S. Varadhan, in *Random Fields, Coll. Math. Soc. Janos Bolyai*, (J. Fritz, J.L. Leibowitz, D. Szasz, eds.), p. 835-873, North-Holland, Amsterdam (1982).
24. I.M. Gelfand and Vilenkin, *Generalized Fuctions*, Vol.IV, Academic Press, New York, London, 1964.
25. N.I. Ahiezer and M. Krein, *Some Questions in the Theory of Moments, Trans. of Math. Monographs, 2*, American Mathematical Society, Rhode Island, 1962.
26. S.G. Mihlin, *Dok. Akad. Nauk. 109*, 701 (1956).
27. M. Avellaneda and A. Majda, in preparation.
28. S.M. Kozlov, Russian Math. Surv. *40* (2), 73 (1985).

Marco Avellaneda
New York University
Courant Institute
251 Mercer Street
New York, NY 10012
U.S.A.

Andrew J. Majda
Program for Computational
and Applied Mathematics
Department of Mathematics
Princeton University
Princeton, NJ 08540
U.S.A.

PROPERTIES OF AVERAGED MODELS OF THE PERIODIC MEDIA MECHANICS

N. Bakhvalov, M. Eglit

Abstract

In homogenization of processes (or stationary states) in periodic structures the equations of other types comparing to original ones often arise. In this work we discuss briefly some of such situations. In particular we consider the problem of conservation of variational properties of a model in homogenization.

Large-scale processes in non-homogeneous media with periodic structure (composites) are often described by the equations of some averaged model,which contain only "averaged" parameters. Properties of averaged models are often not only quantitativly but also qualitativly different from those of components of original non-homogeneous media [2,3,5,8,10].

Here we present a brief survey of results concerning the investigation of the possibility of appearance of new properties or conservation of original properties in the process of averaging for periodic structures.

In the standard homogenization procedure one might apply the multi-scale method. Together with slow variables $x = (x_1,\dots,x_2)$ fast variables $\xi = (\xi_1,\dots,\xi_2)$, $\xi_i = x_i/\varepsilon$ are introduced, where $\varepsilon \ll 1$ is the ratio of the scale of non-homogeneity (period of structure) to the typical scale of the solution of averaged problem, the latter may be taken to be equal to 1. The functions which describe the properties of periodical media are 1-periodic of ξ_i. Let us denote a set of such functions by symbol V. For $f(x,\xi) \in V$ the averaging-over-period operator is defined as

$$<f> = \int_0^1 f(x,\xi)\,d\xi$$

where variables x and ξ are considered to be independent. A solution $u(x,\xi)$, $u = (u_1,\dots,u_m)$, of original equations is written as an asymptotic series

$$u \sim \sum \varepsilon^n v_n(x,\xi) \ , \quad v_n \in V \ .$$

Then a certain procedure is constructed, in the result of which $v_n(x,\xi)$ are determined by some functions $v(x)$ for which "averaged" equations are constructed.

First example of appearance of new properties in the homogenization procedure we get when averaging the equations of periodic elastic medium. The averaged equations (if one does not restrict oneself by an approximation of zero-order) are in fact the equations of some medium, in which stresses depend not only on deformations, but also on high-order derivatives of displacements over coordinates. One of the new properties of such averaged model is the dependence of the propagation velocity of a harmonic wave on its frequency - the dispersion of waves. This phenomenon arises in the composite even if dispersion is absent in all its components.

New properties appear also in averaged models for the periodic visco-elastic media. It is known [1,8] that averaged equations for the system

$$\rho(\xi)\frac{\partial^2 u}{\partial t^2} = \frac{\partial}{\partial x_i}\Big(A_{ij}(\xi)\frac{\partial u}{\partial x_j}\Big) + \frac{\partial}{\partial x_i}\Big(B_{ij}(\xi)\frac{\partial^2 u}{\partial t\partial x_j}\Big), \quad \xi = \frac{x}{\varepsilon}, \quad x = (x_1,\dots,x_s),$$

are of the form

$$<\rho>\frac{\partial^2 v}{\partial t^2} = \frac{\partial}{\partial x_i}\Big(\hat{A}_{ij}\frac{\partial v}{\partial x_j}\Big) + \frac{\partial}{\partial x_i}\Big(\int_0^\infty \hat{B}_{ij}(\tau)\frac{\partial^2 v}{\partial x_j\partial t}\Big|_{t-\tau} d\tau\Big) \ .$$

That is, a system of differential equations is replaced by a system with memory. Analogous systems of equations which contain integrals with respect to time t arise when averaging Maxwell equations [10]:

$$\eta(\xi)\frac{\partial E}{\partial t} + \sigma(\xi)E - \operatorname{rot} H = 0$$

$$\mu(\xi)\frac{\partial H}{\partial t} + \operatorname{rot} E = 0 \ .$$

Estimations of nearness of the solutions of original and averaged problems are given in [9].

Analogous (but nonlinear) systems with memory appear when

studying the problems for the elasto-plastic periodical media [3,5,6] and for the periodical compressible heat-conveying viscous fluids [3,5,7].

The system of equations for these media includes the equations of motion and energy

$$\rho \frac{\partial^2 u^i}{\partial t^2} = \rho F^i + \frac{\partial \sigma^{ij}}{\partial x_j}$$

$$\rho \frac{\partial}{\partial t}\left(e + \frac{u_t^2}{2} \right) = \rho F^i u_t^i + \frac{\partial}{\partial x_j}(\sigma^{ij} u_t^i - q^j)$$

the constituitive equations

$$\frac{\partial \eta^{ij}}{\partial t} = \Phi^{ij}(x^k, \eta^{kl}, \sigma^{kl}, \frac{\partial \sigma^{kl}}{\partial t}, T, \frac{\partial T}{\partial t}, \chi^k)$$

$$\frac{\partial \chi^k}{\partial t} = A^k(x^i, \eta^{ij}, \sigma^{ij}, \frac{\partial \sigma^{ij}}{\partial t}, T, \frac{\partial T}{\partial t}, \chi^k)$$

$$e = e(x^i, \eta^{ij}, \sigma^{ij}, T, \chi^k)$$

$$q^i = q^i(x^k, \eta^{kl}, \sigma^{kl}, T, \frac{\partial T}{\partial x^k}, \chi^k) \ , \quad \rho = \rho(x^i)$$

where u^i, σ^{ij}, $\eta^{ij} = \frac{1}{2}(\partial u^i/\partial x_j + \partial u^j/\partial x_i)$, q^j are components of displacement vector, stress and deformation tensors and heat-flow vector correspondingly, e is an internal energy density, χ^k are the hardening parameters in plasticity. Let Φ^{ij}, A^k, e, q^i, ρ be periodical functions of x^i. Then the averaging procedure leadsto the model where stresses, internal energy density and heat-flow vector are functionals depending on the whole history of deformation and temperature change.

For media in question some different treatment of averaging procedure is proposed in [5,7]. It is possible to receive a set of integro-differential equations for new "averaged" functions, some of which depend on fast variables ξ_i. The number of unknown functions in the new system is in general more than that in the original one. The system includes integration of some of unknown functions over variables ξ_i, but not over time. Contrary to original equations this system does not involve two different scales, L and ε. That is why this system may be treated as an averaged one. Equations of such type correspond to media with additional parameters and additional degrees of freedom.

In particular in the case of layered media such a way gives rise to the models of media with several "phases" present at a point: to each point are assigned several unknown functions (deformations, pressures and so on) corresponding to different phases [3,7].

The example of another type we get when considering the system of integro-differential equations with distributed parameters:

$$R(\xi)\frac{\partial^2 u}{\partial t^2} = \varepsilon^{-s}\int_{-\infty}^{\infty}\dots\int_{-\infty}^{\infty} K(\xi,\lambda)\,u(y)\,dy\ , \quad dy = dy_1\dots dy_s,$$

$\lambda = y/\varepsilon$, $K(\xi,\lambda)$ have the property $K(\xi+m,\lambda+m) = K(\xi,\lambda)$ for every vector $m = (m_1,\dots,m_s)$ with integer m_j. In the homogenization we receive the differential equation of infinite order [5]:

$$\hat{L}v \sim \sum_{q+l\geq 2} \varepsilon^{q+l-2} h_{qj_1\dots j_l} \frac{\partial^{q+l} v}{\partial t^q \partial x_{j_1}\dots\partial x_{j_l}} \sim 0,\ h_{qj_1\dots j_l} = \text{const}.$$

Analogous averaged equations arise when averaging the systems of operator equations with periodic properties [2,5]. This case involves integro-differential equations mentioned above, standard differential equations for periodic structures, equations defined on periodic systems of discrete points as particular cases. The equations have the form

$$L_0\frac{\partial^2 u}{\partial t^2} - L_1 u = 0$$

where L_0, L_1 are periodic operators:

$$L_0 S_{\vec{m}} u \equiv S_{\vec{m}} L u\ ,$$

$\vec{m} = (m_1,\dots,m_s)$ is an arbitrary vector with integer coordinates; $S_{\vec{m}}$ is the translation operator

$$S_{\vec{m}} f(\xi) = f(x + \vec{m}\varepsilon)\ .$$

Let us search a solution of averaged differential equation of infinte order $\hat{L}v \sim 0$ in the form $v = e^{i}(k_j x_j - \omega t)$. Substituting it into the equation we obtain

$$\sum_{q+l\geq 2} \varepsilon^{q+l-2} h_{qj_1\ldots j_l} (-i\omega)^q (ik_{j_1})\ldots(ik_{j_l}) \sim 0 .$$

This relation shows that the wave velocities depend on their frequencies; such a phenomenon is called the wave dispersion.

One of the well-known effects in the mechanics of heterogeneous media is the effect of anomalous decrease of the sound velocity in mixtures. For instance the value of the sound velocity in air is approximately **330 m/s**, in water - **1400 m/s**, and in mixture of water and air bubbles it may be equal to **10 m/s**. Let us consider a system of equations of the elasticity theory for periodic media which consists of isotropic homogeneous elastic media:

$$\rho(\xi)\frac{\partial^2 u_i}{\partial t^2} = \frac{\partial}{\partial x_i}(\lambda(\xi)\operatorname{div}\vec{u}) + \frac{\partial}{\partial x_j}(\mu(\xi)(\frac{\partial u_i}{\partial x_j} + \frac{\partial u_j}{\partial x_i}))$$

$$\rho > 0 , \quad \lambda > 0 , \quad \mu > 0 .$$

Standard averaging procedure yield in zero-order approximation of the system of hyperbolic type

$$<\rho> \frac{\partial^2 \bar{v}}{\partial t^2} = \frac{\partial}{\partial x_i}(\hat{A}_{ij} \frac{\partial \bar{v}}{\partial x_j}) .$$

The value of the velocity of a small perturbation in media described by such a system may depend on propagation direction. Let us consider the medium where one of the components fills up spheres with centers in the points $(n_1\varepsilon, n_2\varepsilon, n_3\varepsilon)$ and the other fills up remaining space. It may be shown [1] that when $\mu \to 0$ the perturbation velocities for averaded systems tend to zero or tend to $\bar{c} = (<\rho><\lambda^{-1}>)^{-1}$. If volumes of each phase are equal, the density and the compressibility of the phases are equal to those of air and water correspondingly then the value of $\bar{c}$ is essentially less than the sound velocities both in air and water.

If $\mu = 0$ from the beginning then the perturbation velocity differs from $\bar{c}$; quantitative difference is not large for the medium in consideration; in the case when spheres are filled by water and remaining space is filled by air quantitative difference between small perturbation velocity in averaged media and $\bar{c}$ is very large.

Many problems in mechanics are described by equations which are the Euler equations for some functional. Of interest is the question whether it is possible to construct homogenized equations for approximation of any order on ε such that they are also the Euler equations for some functional. The answer to this question allows to clarify whether it is possible for a homogenized model to be dissipative when there is no dissipation in components of the non-homogeneous medium. Now we shall formulate some results from [4] concerning the problems mentioned above for periodic media. We shall study the problem of finding the stationary points of the functional

(1) $I(u) = \int U(x, \xi, \text{grad } u) dx$,

where U is the 1-periodic over each variable $\xi_j \equiv x_j/\varepsilon$ function and the integration is performed over all s-dimensional space. It is assumed that the function U is infinitely differentiable over arguments x_j, $u_j = \partial u/\partial x_j$, the derivatives of any order over these arguments being uniformly bounded. If u is the stationary point of functional $I(u)$, then

(2) $\Lambda(u, \varphi) = \int \dfrac{\partial U}{\partial u_j} \dfrac{\partial \varphi}{\partial x_j} dx = 0$

for any function φ such that $u + \tau\varphi$ at sufficiently small τ belongs to the region of definiteness of functional I. Let us denote by C the space of 1-periodic over ξ functions $\varphi(x, \xi)$ defined for all x, equal to zero at $(x, x) \geqslant R^2$, $R < \infty$ and such that each of the derivatives over x is uniformly bounded over x and ξ . In relation (2) we use the trial functions $\varphi \in C$. For functions u satisfying the condition $\Lambda(u, \varphi) = 0$ $\forall \varphi \in C$, functional $I(u)$ may be not definite (the right-hand side integral diverges). So, below when saying "the stationary point of functional $I(u)$" we mean only that

(3) $\Lambda(u, \varphi) \sim 0$ $\forall \varphi \in C$.

Theorem 1. It is possible to construct an asymptotic expansion of the solution of problem (3)

$$(4) \quad u \sim v(x) + \varepsilon v_1(x,\xi) + \dots + \varepsilon^n v_n(x,\xi) + \dots$$

such that

$$\Lambda(u,\varphi) \sim \int \hat{L}(v)\langle\varphi\rangle dx \quad \forall\varphi \in C$$

and $v_n(x,\xi)$ $\forall n > 0$ are functions of x, ξ, $v(x)$ and derivatives of v over x of order not exceeding n; $\hat{L}(v)$ is an asymptotic series in powers of ε with coefficients depending on x, v and its derivatives over x.

Equation

$$(5) \quad \hat{L}(v) \sim 0$$

may be considered as the averaged equation for the Euler equation corresponding to the stationary point of functional (1).

Asymptotic expansion (4) satisfying Theorem 1 is not unique: the functions v_n are determined up to arbitrary functions independent of ξ. It is possible to choose these functions so that $\langle v_n\rangle \equiv 0$.

Theorem 2. If all $\langle v_n\rangle \equiv 0$ $\forall n > 0$, equation (5) is the Euler equation for functional

$$\hat{I}(v) = \int \hat{U}(v)dx\ , \quad \hat{U}(v) \sim \langle U(x,\xi,\text{grad}\ u(v))\rangle\ ,$$

where $u(v)$ is determined by series (4).

Note that the fulfillment of the relation

$$\int \hat{L}(v)\langle\varphi\rangle dx \sim 0$$

only for $\varphi \in C$ is often sufficient for obtaining strict mathematical results about nearness of the solutions of original and averaged problems.

Let us now consider a functional describing dynamic processes

$$(6) \quad I(u) = \int_t \int_x (T - U)dx\,dt\ ,$$

where $T = \rho(x,\xi)u_t^2/2$, $U = U(x,\xi,\text{grad}\ u)$ posesses the same properties as before. Using the notations $x = (x_1,\dots,x_{s+1})$, $x_{s+1} \equiv t$,

$\xi = (\xi_1,\dots,\xi_{s+1}) = x/\varepsilon$ it may be seen that functional (6) belongs to the class the functionals considered in Theorems 1 and 2, so these theorems are valid for dynamical problems. The functional for averaged problem is

$$\hat{I}(v) = \int_t \int_x (\hat{T} - \hat{U})\,dx\,dt\ ,$$

where $\hat{T}(v) = \langle \rho(x,\xi)\cdot(u(v))_t^2/2\rangle$ and $\hat{U}(v) = \langle U(x,\xi,\mathrm{grad}\ u(v))\rangle$. In the approximation of any order on ε there exists the energy integral

$$\int (\hat{T} + \hat{U})\,dx \sim \text{const}\ .$$

Expression $\hat{T}(v)$ may be called the averaged kinetic energy density and $\hat{U}(v)$ - the averaged potential energy density. Note, that $\hat{T}$ and $\hat{V}$ depend on the arbitrary high order derivatives of function v over coordinates and time.

Theorem 3. There exists the transformation

$$v \sim w + \sum_{n=1}^{\infty} \varepsilon^n w_n(w)$$

in the class of finite functions such that functional $\hat{I}(v(w))$ involve the derivatives of w with time differentiation no higher than the first order.

For linear media $(U = a_{ij}\ \partial u/\partial x_i\ \partial u/\partial x_j)$ it is possible to introduce an unlimited number of expressions $\bar{T}(\omega)$ and $\bar{U}(\omega)$, such that

$$(7)\quad \hat{I} \sim \iint (\bar{T}(\omega) - \bar{U}(\omega))\,dx\,dt\ ,\quad \int (\bar{T}(\omega) + \bar{U}(\omega))\,dx \sim \text{const},$$

ω is a new variable:

$$v \sim \omega + \sum_{q+l>0} \varepsilon^{q+l} g_{qj_1\dots j_l} \frac{\partial^{q+l} v}{\partial t^q \partial x_{j_1}\dots\partial x_{j_l}}\ ,\quad g_{qj_1\dots j_l} = \text{const}\ .$$

As a rule

$$\hat{T}(v(\omega)) \not\sim \bar{T}(\omega)\ ,\quad \hat{V}(v(\omega)) \not\sim \bar{V}(\omega)$$

though

$$\hat{T}(v(\omega)) - \hat{V}(v(\omega)) \sim \bar{T}(\omega) - \bar{V}(\omega)\ .$$

Theorem 4. Among all the possible ways to write (7) for linear media there exist two canonic forms, (A) and (B):

(A) $\bar{T}(\omega) = \langle\rho\rangle\omega_t^2/2$, $\bar{U}(\omega)$ is quadratic form of derivatives $\partial^m\omega/\partial x_{j1}\ldots\partial x_{jm}$, $1 \leq m < \infty$; coefficients of the form depend not only on a_{ij}, but also on ρ.

(B) $\bar{T}(\omega)$ is quadratic form of derivatives $\partial^{m+1}\omega/\partial t\partial x_{j1}\ldots\partial x_{jm}$, $1 \leq m < \infty$; $\bar{U}(\omega)$ have the same form as in (A), but $\bar{U}(\omega)$ is determined only by a_{ij}, $\bar{T}(\omega)$ depends not only on ρ but also on a_{ij}.

In [4] functionals of more general type are also considered, when in (1)

(8) $u = (u^1,\ldots,u^r)$, $U = U(\varepsilon, x, \xi, \text{grad } u, \text{grad}^2 u, \ldots)$.

One comes across functionals of this type when components of nonhomogeneous periodic media are described by equations of moment elasticity theory or when these components are periodic structures themselves, Theorems 1-3 remain valid also for the case (8).

References

[1] N. Bakhvalov, "On the sound velocity in mixtures", DAN SSSR, **245** (1979), 1345-1348, (in Russian).

[2] N. Bakhvalov, "Homogenization and perturbation problems", Computing Methods in Aplied Science and Engineering, Amsterdam, North Holland, 1980, 645-658.

[3] N. Bakhvalov, M.Eglit, "Processes in periodic media which could not be described in terms of averaged characteristics", DAN SSSR, **268** (1983), 836-840, (in Russian).

[4] N. Bakhvalov, M. Eglit, "Variational properties of averaged equations of periodic media", Trudy MIAN, Moscow, **192** (1990), 5-19, (in Russian).

[5] N. Bakhvalov, G. Panasenko, "Homogenization: averaging processes in periodic media", Mathematical Problems in the Mechanics of Composite Materials, Kluwer Academic Publishers, Dordrecht-Boston-London, 1989.

[6] M. Eglit, "On the averaged description of the processes in periodic elastic-plastic media", MKM, Riga, **5** (1984), 825-831, (in Russian).

[7] M. Eglit, "On averaged description of large-scale processes in periodic viscous compressable media", Mechanics. Current Problems, MGU, Moscow, 1987, 121-126, (in Russian).

[8] J.L. Lions, "Some methods in the mathematical analysis of systems and their control", Science Press, Beijing, China, 1981.

[9] S. Mozolin, "About nearness of solutions of original and averaged problems of electrodynamics and visco-elasticity", DAN SSSR, **273** (1983), 330-333, (in Russian).

[10] E. Sanches-Palencia, "Non-homogenious media and vibration theory", Lecture Notes in Physics, 127, Berlin, Springer-Verlag, 1980.

Nickolaj S. Bakhvalov
Dept. of Numerical Mathematics
USSR Academy of Sciences
MOSCOW, 117901, USSR

Margarita E. Eglit
Dept. of Mathematics and
Mechanics, Moscow State Univ.
MOSCOW, 119899, USSR

HOMOGENIZATION OF A CLASS OF STOCHASTIC PARTIAL DIFFERENTIAL EQUATIONS

A. Bensoussan

INTRODUCTION

It is well known that the classical homogenization theory of elliptic and parabolic equations can be derived by probabilistic methods, through the limit of diffusion stochastic processes (see A. BENSOUSSAN - J.L. LIONS - G. PAPANICOLAOU [4]).

This motivates the study of stochastic P.D.E. instead of stochastic ordinary differential equations, with oscillatory coefficients. In this case, the probabilistic approach is the unique one possible. Indeed, the counter part of elliptic and parabolic equations arising in the context of stochastic O.D.E. would lead to infinite dimensional P.D.E. and the corresponding theory is not very advanced. Moreover stochastic P.D.E. with oscillatory coefficients arise also in a natural fashion in the theory of nonlinear filtering (see A. BENSOUSSAN - G. BLANKENSHIP [3]) and it is important to understand the behaviour limit of the solution. We consider here a class of nonlinear stochastic P.D.E., recently introduced by N. NAGASE [6] (See also A. BENSOUSSAN [1]) involving just continuous coefficients. Some stability results with respect to perturbation of coefficients are proved in N. NAGASE [6]. Our results are in the spirit of those of NAGASE.

1. SETTING OF THE PROBLEM

1.1. Notation - Assumptions

Let $\mathcal{O}$ be a smooth bounded domain of $\mathbb{R}^n$. We consider a family of matrices $a^\varepsilon(x) \in L(\mathbb{R}^n; \mathbb{R}^n)$ satisfying

(1.1) $a^\varepsilon(x)$ is measurable and

$$a^\varepsilon(x)\xi \cdot \xi \geq \alpha|\xi|^2, \quad (a^\varepsilon(x))^{-1}\xi \cdot \xi \geq \beta|\xi|^2$$

$$\alpha, \beta > 0.$$

The classical assumptions of homogenization theory (L. TARTAR [10]) are the following : there exists a sequence of vectors $v^\varepsilon(x) \in \mathbb{R}^n$ such that

(1.2) i) $v^\varepsilon - x \to 0$ in $H^1(\mathcal{O}; \mathbb{R}^n)$ weakly

ii) $a^\varepsilon (Dv^\varepsilon)^* \to a$ in $L^2(\mathcal{O}; L(\mathbb{R}^n; \mathbb{R}^n))$ weakly

where $(Dv^\varepsilon)_{ij} = \dfrac{\partial v_i^\varepsilon}{\partial x_j}$.

iii) $\text{div}\,(a^\varepsilon (Dv^\varepsilon)^* \xi) \to \text{div}\,(a\xi)$ in $H^{-1}(\mathcal{O})$ strongly $\forall \xi \in \mathbb{R}^n$.

iv) $\|Dv^\varepsilon\| \leq c.$

The following important facts are derived from homogenization theory (see A. BENSOUSSAN, J.L. LIONS, G. PAPANICOLAOU [4], L. TARTAR [10],...) or the G-convergence theory (E. DE GIORGI, S. SPAGNOLO [5])

(1.3) $a(x)$ satisfies the properties (1.1) with the same constants α, β

(1.4)
$$\begin{aligned}
&\text{Let } f^\varepsilon \in H^{-1}(\mathcal{O}), f^\varepsilon \to f \text{ in } H^{-1}(\mathcal{O}) \text{ strongly ;}\\
&\text{and let } z^\varepsilon \text{ be the solution of}\\
&\qquad A^\varepsilon z^\varepsilon = f^\varepsilon \qquad z^\varepsilon \in H_0^1(\mathcal{O})\\
&\text{where } A^\varepsilon = -\operatorname{div}(a^\varepsilon D\cdot)\\
&\text{then } z^\varepsilon \to z \text{ in } H_0^1(\mathcal{O}) \text{ weakly}\\
&\qquad a^\varepsilon D z^\varepsilon \to aDz \text{ in } L^2(\mathcal{O};\mathbb{R}^n)) \text{ weakly}\\
&\qquad Dz^\varepsilon - (Dv^\varepsilon)^* Dz \to 0 \qquad L^2(\mathcal{O};\mathbb{R}^n) \text{ strongly}\\
&\qquad Az = f, \qquad A = -\operatorname{div}(aD\cdot).
\end{aligned}$$

Note also that :

(1.5)
$$\begin{aligned}
&\text{if } g^\varepsilon \in H^{-1}(\mathcal{O}), \qquad g^\varepsilon \to g \text{ in } H^{-1}(\mathcal{O}) \text{ strongly}\\
&\text{and } (A^\varepsilon)^* \zeta^\varepsilon = g^\varepsilon, \qquad \zeta^\varepsilon \in H_0^1(\mathcal{O})\\
&\text{with } (A^\varepsilon)^* = -\operatorname{div}((a^\varepsilon)^* D\cdot)\\
&\text{then } \zeta^\varepsilon \to \zeta \text{ in } H_0^1(\mathcal{O}) \text{ weakly with}\\
&\qquad A^*\zeta = g.
\end{aligned}$$

We shall set in the sequel

(1.6)
$$H = L^2(\mathcal{O}), \qquad V = H_0^1(\mathcal{O})$$

Let next $g^\varepsilon(t,u), B^\varepsilon(t,u)$ be such that :

(1.7)
$$\begin{aligned}
&i)\ g^\varepsilon(t,u) : (O,T) \times H \to H \quad \textit{measurable}\\
&ii)\ \forall \delta, \exists \eta(\delta) \text{ such that}\\
&\qquad |u-v| < \eta \Rightarrow |g^\varepsilon(t,u) - g^\varepsilon(t,v)| \leq \delta\\
&\qquad \text{(uniform continuity with respect to } u).\\
&iii)\ |g^\varepsilon(t,u)|_H \leq g(1+|u|)
\end{aligned}$$

(1.8)
$$\begin{aligned}
&i)\ B^\varepsilon(t,u) : (O,T) \times H \to H^m \text{ measurable}\\
&ii)\ \forall \delta, \exists \eta(\delta) \text{ such that}\\
&\qquad |u-v| < \eta \Rightarrow |B^\varepsilon(t,u) - B^\varepsilon(t,v)|_{H^m} \leq \delta\\
&iii)\ |B^\varepsilon(t,u)|^2 \leq b(1+|u|^2)\\
&iv)\ |B^\varepsilon(t,u) - B^\varepsilon(s,u)|^2 \leq 0(t-s)(|u|^2+1)
\end{aligned}$$

with $0(h)$ monotone increasing, $0(h) \to 0$, as $h \to 0$.

We also assume the following convergence properties :

(1.9) $g^\varepsilon(t,u) \to g(t,u)$ in H a.e.t, and $\forall u \in H$.
(1.10) $B^\varepsilon(t,u) \to B(t,u)$ in H^m a.e.t, and $\forall u \in H$

and it is obvious that $g(t,u), B(t,u)$ satisfy the same properties (1.7), (1.8).

1.1. The problem

We consider the stochastic P.D.E. for fixed ε,

$$(1.11) \qquad \begin{aligned} & dy^\varepsilon + A^\varepsilon y^\varepsilon dt = g^\varepsilon(t, y^\varepsilon(t))dt + B^\varepsilon(t, y^\varepsilon(t)) \cdot dw^\varepsilon \\ & y^\varepsilon(0) = y_0 \end{aligned}$$

where

$$(1.12) \qquad y_0 \in H.$$

Following N. NAGASE [6], A. BENSOUSSAN [1], we call a solution of (1.11) on $[0,T]$ a system

$$\Omega^\varepsilon, \mathcal{A}^\varepsilon, P^\varepsilon, \mathcal{F}^{t,\varepsilon}, w^\varepsilon(t), y^\varepsilon(t)$$

such that

(1.13)$\Omega^\varepsilon, \mathcal{A}^\varepsilon, P^\varepsilon$, is a probability space , $\mathcal{F}^{t,\varepsilon}$a filtration,
$w^\varepsilon(t)$is a $\mathbb{R}^m$ standard Wiener process with respect to
$\mathcal{F}^{t,\varepsilon}, y^\varepsilon(.)$ is an element of $L^2_{\mathcal{F}^{t,\varepsilon}}(0,T;V) \cap L^2(\Omega^\varepsilon, \mathcal{A}^\varepsilon, P^\varepsilon; C(0,T;H))$

and

$$(1.14) \qquad \text{a.s. } y^\varepsilon(t) + \int_0^t A^\varepsilon y^\varepsilon(s)ds = y_0 + \int_0^t g^\varepsilon(s, y^\varepsilon(s))ds + \\ + \int_0^t B^\varepsilon(s, y^\varepsilon(s)) \cdot dw^\varepsilon, \qquad \forall t \in [0,T]$$

We proved in the paper mentionned above that there exists a solution of (1.11), in the sense of (1.13), (1.14).

In a similar way, we shall consider the limit equation :

$$(1.15) \qquad \begin{aligned} & dy + Ay\ dt = g(t, y(t))dt + B(t, y(t)) \cdot dw \\ & y(0) = y_0 \end{aligned}$$

which should be interpreted again in the sense (1.13), (1.14), and has solutions according to the same result.

Let $S = C(0,T;\mathbb{R}^m) \times L^2(0,T;H)$, our objective is to prove the following

Theorem 1.1. *We assume (1.1), (1.2), (1.7), (1.8), (1.9), (1.10), then there exists a sequence ε_j and a solution of (1.15) such that :*

$$w_{\varepsilon_j}, y_{\varepsilon_j} \to w, y \text{ in law as random variables with values in } S. \tag{1.16}$$

2. ESTIMATES

We begin by proving a priori estimates.

2.1. First estimates

We shall prove the

Lemma 2.1. *We have :*

$$E^\varepsilon \int_0^T \|y^\varepsilon(t)\|^2 dt \leq C \tag{2.1}$$

$$E^\varepsilon |y^\varepsilon(t)|^2, \qquad E^\varepsilon |y^\varepsilon(t)|^4 \leq C, \qquad \forall t \in [0,T]. \tag{2.2}$$

Proof.

We write the energy equality :

$$\begin{aligned} E^\varepsilon |y^\varepsilon(t)|^2 + 2E^\varepsilon \int_0^t < A^\varepsilon y^\varepsilon, y^\varepsilon > ds = |y^0|^2 + 2E^\varepsilon \int_0^t (g(y^\varepsilon), y^\varepsilon) ds \\ + E^\varepsilon \int_0^t |B^\varepsilon(y^\varepsilon)|^2 ds \end{aligned} \tag{2.3}$$

hence

$$\begin{aligned} E^\varepsilon |y^\varepsilon(t)|^2 + 2\alpha E^\varepsilon \int_0^t \|y^\varepsilon(s)\|^2 ds \leq |y^0|^2 + 2g \quad E^\varepsilon \int_0^t (1 + |y^\varepsilon|)|y^\varepsilon| ds \\ + bE^\varepsilon \int_0^t (1 + |y^\varepsilon|^2) ds. \end{aligned}$$

This yields easily making use of Gronwall's inequality, to (2.1) and the 1st estimate (2.2).

To derive the 2nd estimate (2.2), we first notice the stochastic energy equality

$$\begin{aligned} d|y^\varepsilon(t)|^2 + 2 < A^\varepsilon y^\varepsilon, y^\varepsilon > dt = (2 < g^\varepsilon(y^\varepsilon), y^\varepsilon > + |B^\varepsilon(y^\varepsilon)|^2) dt \\ + 2(y^\varepsilon, B^\varepsilon(y^\varepsilon)) \cdot dw^\varepsilon \end{aligned}$$

hence

$$\begin{aligned} d|y^\varepsilon(t)|^4 + 4|y^\varepsilon(t)|^2 < A^\varepsilon y^\varepsilon, y^\varepsilon > dt = (2|y^\varepsilon(t)|^2(2(g^\varepsilon(y^\varepsilon), y^\varepsilon) \\ + |B^\varepsilon(y^\varepsilon)|^2) dt + 4|(y^\varepsilon, B^\varepsilon(y^\varepsilon))|^2 dt + 4|(y^\varepsilon, (t)|^2 (y^\varepsilon, B^\varepsilon(y^\varepsilon)) \cdot dw^\varepsilon \end{aligned}$$

It follows that :

$$\begin{aligned} \frac{d}{dt} E^\varepsilon |y^\varepsilon(t)|^4 + 4\alpha E |y^\varepsilon(t)|^2 \|y^\varepsilon(t)\|^2 \leq 4|y^\varepsilon(t)|^2 (1 + |y^\varepsilon(t)|)|y^\varepsilon(t)| \\ + 2b|y^\varepsilon(t)|^2 (1 + |y^\varepsilon(t)|^2) \end{aligned}$$

and again this inequality implies easily the 2nd estimate (2.2). □

We can also improve the 1st estimate (2.2) as follows

Lemma 2.2. *We have the estimate*

$$E^\varepsilon \sup_{t\in[0,T]} |y^\varepsilon(t)|^2 \leq C. \tag{2.4}$$

Proof.
From the proof of Lemma 2.1, we have obtained

$$|y^\varepsilon(t)|^2 \leq C + C\int_0^t |y^\varepsilon(s)|^2 ds + 2\int_0^t (y^\varepsilon, B^\varepsilon(y^\varepsilon)) \cdot dw^\varepsilon$$

hence using the 1st estimate (2.2)

$$\begin{aligned} E^\varepsilon \sup_{t\in[0,T]} |y^\varepsilon(t)|^2 &\leq C + 2E^\varepsilon \sup_{t\in[0,T]} \Big| \int_0^t (y^\varepsilon, B^\varepsilon(y^\varepsilon)) \cdot dw^\varepsilon \Big| \\ &\leq C + (E^\varepsilon \int_0^T |y^\varepsilon(t)|^4 dt)^{\frac{1}{2}} \\ &\leq C \end{aligned}$$

according to the 2nd estimate (2.2). □

2.2. Additional estimate

We prove now the following important estimate :

Lemma 2.3. *One has the property*

$$E^\varepsilon \sup_{|\theta|\leq\delta} \int_0^T \|y^\varepsilon(t+\theta) - y^\varepsilon(t)\|^2_{V^*} dt \leq C\delta, \qquad \forall\delta < 1$$

Proof.
We extend $y^\varepsilon(t)$ by 0 outside the interval $(0, T)$. We may assume $\theta > 0$. Define :

$$I^\varepsilon = E^\varepsilon \sup_{0\leq\theta\leq\delta} \int_0^T \|y^\varepsilon(t+\theta) - y^\varepsilon(t)\|^2_{V^*} dt \leq I_1^\varepsilon + I_2^\varepsilon$$

where

$$\begin{aligned} I_1^\varepsilon &= E^\varepsilon \sup_{0\leq\theta\leq\delta} \int_0^{T-\delta} \|y^\varepsilon(t+\theta) - y^\varepsilon(t)\|^2_{V^*} dt \\ I_2^\varepsilon &= E^\varepsilon \sup_{0\leq\theta\leq\delta} \int_{T-\delta}^T \|y^\varepsilon(t+\theta) - y^\varepsilon(t)\|^2_{V^*} dt \\ &\leq C \quad E^\varepsilon \sup_{0\leq\theta\leq\delta} \int_{T-\delta}^T |y^\varepsilon(t+\theta) - y^\varepsilon(t)|^2_H dt \\ &\leq C\delta, \text{ thanks to (2.4).} \end{aligned}$$

Consider now I_1^ε. We have :

$$E^\varepsilon \sup_{0\leq\theta\leq\delta} \int_0^{T-\delta} \| \int_t^{t+\theta} A^\varepsilon y^\varepsilon ds\|_{V^*}^2 dt \leq C\delta$$

Similarly

$$E^\varepsilon \sup_{0\leq\theta\leq\delta} \int_0^{T-\delta} \| \int_t^{t+\theta} g(y^\varepsilon) ds\|_{V^*}^2 dt \leq C\delta^2.$$

Finally

$$E^\varepsilon \sup_{0\leq\theta\leq\delta} \int_0^{T-\delta} \| \int_t^{t+\theta} B^\varepsilon(y^\varepsilon)\cdot dw^\varepsilon\|_{V^*}^2 dt \leq$$

$$C \quad E^\varepsilon \sup_{0\leq\theta\leq\delta} \int_0^{T-\delta} | \int_t^{t+\theta} B^\varepsilon(y^\varepsilon)\cdot dw^\varepsilon|^2 dt \leq$$

$$C\int_0^{T-\delta} (E^\varepsilon \sup_{0\leq\theta\leq\delta} | \int_t^{t+\theta} B^\varepsilon(y^\varepsilon)\cdot dw^\varepsilon|^2) dt \leq$$

$$C\int_0^{T-\delta} (E^\varepsilon \int_t^{t+\theta} |B^\varepsilon(y^\varepsilon)|^2 dt) dt \leq C\delta$$

and thus the desired result (2.5) have been proved. □

3. CONVERGENCE

3.1. Tightness property

In the Hilbert space $L^2(0,T;H)$ we consider a set Z depending on 3 constants K, L, M and two sequences μ_n, ν_n of numbers such that $\mu_n, \nu_n \geq 0$ and $\mu_n, \nu_n \to 0$ as $n \to \infty$. The set Z is defined as follows :

$$Z = \{z | \int_0^T \|z(t)\|_V^2 dt \leq K, \qquad |z(t)|_H^2 \leq L \text{ a.e.t.}, \quad \sup_{|\theta|\leq\mu_n} \int_0^T \|z(t+\theta) - z(t)\|_{V^*}^2 dt \leq \nu_n M, \forall n\} \tag{3.1}$$

where z is extended by 0 outside (0,T).

The set Z is a *compact subset* of $L^2(0,T;H)$ (see A. BENSOUSSAN [1]). Let $S = C(0,T;\mathbb{R}^m) \times L^2(0,T;H)$. We define on S the probability measure π^ε image on S of the probability P^ε on Ω^ε by the map

$$\omega \in \Omega^\varepsilon \Rightarrow w_\varepsilon(\cdot), y_\varepsilon(\cdot)$$

We shall prove the :

Lemma 3.1. *The family π^ε is uniformly tight.*

Proof.

We must prove that $\forall\delta, \exists W_\delta \times Z_\delta$ a compact subset of S such that :

$$\pi^\varepsilon(W_\delta \times Z_\delta) \geq 1-\delta \qquad \forall\varepsilon. \tag{3.2}$$

A generic point of S is denoted $b(\cdot), z(\cdot)$. Let

$$W_\delta = \{b(\cdot)| \sup_{t\in[0,T]} |b(t)| \leq q_\delta, \quad \sup_{\substack{t_1,t_2\in[0,T]\\ |t_1-t_2|<\frac{1}{n^6}}} |b(t_2 - b(t_1)| < \frac{r_\delta}{n}, \quad \forall n\}$$

and Z_δ be a set of the form (3.1) with sequences μ_n, ν_n tending to 0 and satisfying:

$$\sum_n \frac{\mu_n}{\nu_n} < \infty$$

for instance pick $\nu_n = \frac{1}{n}, \mu_n = \frac{1}{n^3}$.

Denote by $K_\delta, L_\delta, M_\delta$ the constants entering into the definition of Z, to be chosen in function of δ. We have

$$\pi_\varepsilon(b(\cdot), z(\cdot) \notin W_\delta \times Z_\delta) \leq P^\varepsilon[\{\sup_{t\in[0,T]} |w_\varepsilon(t)| > q_\delta\} \bigcup_n$$

$$\{\sup_{\substack{|t_1-t_2|<\frac{1}{n^6}\\ t_1,t_2\in[0,T]}} |w_\varepsilon(t_2) - w_\varepsilon(t_1)| > \frac{r_\delta}{n}\} \bigcup \{\int_0^T \|y_\varepsilon(t)\|^2 dt > K_\delta\}$$

$$\bigcup \{\sup_{0\leq t\leq T} |y_\varepsilon(t)|^2 > L_\delta\} \bigcup_n \{\sup_{|\theta|\leq\mu_n} \int_0^T \|y_\varepsilon(t+\theta) - y_\varepsilon(t)\|_{V^*}^2 dt > \nu_n M_\delta\}]$$

$$\leq \frac{\sqrt{T}}{q\delta} + \sum_n (\frac{2n}{r_\delta})^4 \sum_{i=0}^{[Tn^6]} E^\varepsilon \sup_{t\in[\frac{i}{n^6},\frac{i+1}{n^6}]} |w_\varepsilon(t) - w_\varepsilon(\frac{i}{n^6})|^4$$

$$+ \frac{1}{K_\delta} E^\varepsilon \int_0^T \|y_\varepsilon(t)\|^2 dt + \frac{1}{L_\delta} E^\varepsilon \sup_{t\in[0,T]} |y^\varepsilon(t)|^2$$

$$+ \sum_n \frac{1}{\nu_n M_\delta} E^\varepsilon \sup_{|\theta|\leq\mu_n} \int_0^T \|y_\varepsilon(t+\theta) - y_\varepsilon(t)\|_{V^*}^2 dt$$

and from Lemma 2.1, 2.2, 2.3

$$\leq C(\frac{\sqrt{T}}{q_\delta} + \frac{1}{r_\delta^4} + \frac{1}{K_\delta} + \frac{1}{L_\delta} + \frac{1}{M_\delta}) \leq \delta$$

for a convenient choice of $q_\delta, r_\delta, K_\delta, L_\delta, M_\delta$. The result (3.2) follows. □

3.2 Proof of Theorem 3.1

According to Prokhorov's theorem [8], the family π^ε is relatively compact in the set $\mathcal{P}(S)$ of probability measures on S, equipped with the weak convergence topology. Therefore, we can extract a subsequence, denoted π^{ε_j} such that :

$$\pi^{\varepsilon_j} \to \pi, \qquad \text{i.e.,} \qquad \forall\phi : S \to \mathbb{R}, \tag{3.3}$$

continuous and bounded

$$\int \phi(b(\cdot), z(\cdot))d\pi^{\varepsilon_j} \to \int \phi(b(\cdot), z(\cdot))d\pi.$$

By Skorokhod's theorem [9], there exists a probability space $\Omega, \mathcal{A}, \mathcal{P}$, and random variables, $w(\cdot,\omega), y(\cdot,\omega)$, $b_{\varepsilon_j}(\cdot,\omega)$, $z_{\varepsilon_j}(\cdot,\omega)$ on $(\Omega, \mathcal{A}, \mathcal{P})$ with values in S, such that

(3.4) The probability law of $b_{\varepsilon_j}(\cdot), z_{\varepsilon_j}(\cdot)$ is π^{ε_j}

(3.5) $b_{\varepsilon_j}(\cdot,\omega), z_{\varepsilon_j}(\cdot,\omega) \to w(\cdot,\omega), y(\cdot,\omega)$ a.s. as $j \to \infty$

(3.6) the probability law of $(w(\cdot), y(\cdot))$ is π.

Define $\mathcal{G}^{t,\varepsilon_j} = \sigma(1\!\!1_t b_{\varepsilon_j}(\cdot,\omega)), 1\!\!1_t z_{\varepsilon_j}(\cdot,\omega))$. We can first check that $b_{\varepsilon_j}(t;\omega)$ is a $\mathcal{G}^{t,\varepsilon_j}$ Wiener process.

The 2nd step consists in showing that an equation similar to (1.11) holds, namely

$$\begin{aligned} &dz_{\varepsilon_j} + A^{\varepsilon_j} z_{\varepsilon_j} dt = g^{\varepsilon_j}(t, z_{\varepsilon_j}(t))dt + B^{\varepsilon_j}(t, z_{\varepsilon_j}(t)) \cdot db^{\varepsilon_j} \\ &z_{\varepsilon_j}(0) = y_0. \end{aligned} \tag{3.7}$$

Indeed we discretize the time and define $k = \dfrac{T}{N+1}$. We define a deterministic map $\phi^{k,\varepsilon}$ from $S \to L^2(0,T;H)$ by setting for any $b(\cdot), z(\cdot)$ in S,

$$\zeta^{k,\varepsilon}(\cdot) = \phi^{k,\varepsilon}(b(\cdot), z(\cdot))$$

$$\zeta^{k,\varepsilon}(t) + \int_0^t A^\varepsilon \zeta^{k,\varepsilon} ds = \int_0^t g^\varepsilon(t, z(s))ds + \int_0^{k[\frac{t}{k}]} \chi_k^\varepsilon(s)db(s) + y_0$$

where

$$\chi_k^\varepsilon(t) = B^\varepsilon(rk, \overline{z_k^r}), t \in [rk, (r+1)k[$$

$$\overline{z_k^r} = \begin{cases} \frac{1}{k}\int_{(r-1)k}^{rk} z(t)dt & r = 1 \dots N+1 \\ y_0 & r = 0 \end{cases}$$

Note that $\phi^{k,\varepsilon}$ is continuous.

Define

$$\begin{aligned} z^{k,\varepsilon_j}(\cdot,\omega) &= \phi^{k,\varepsilon_j}(b_{\varepsilon_j}(\cdot), z_{\varepsilon_j}(\cdot)) \\ y^{k,\varepsilon_j}(\cdot,\omega) &= \phi^{k,\varepsilon_j}(w_{\varepsilon_j}(\cdot), y_{\varepsilon_j}(\cdot)). \end{aligned}$$

We first check that

$$E^{\varepsilon_j} \int_0^T |y_{\varepsilon_j}(t) - y^{k,\varepsilon_j}(t)|^2 dt \to 0 \text{as } k \to 0, \varepsilon_j \text{fixed} \tag{3.8}$$

This follows from classical arguments of numerical analysis.

Define finally $\hat{z}_{\varepsilon_j}(t)$ by setting

$$d\hat{z}_{\varepsilon_j} + A^{\varepsilon_j}\hat{z}_{\varepsilon_j}(t)dt = g^{\varepsilon_j}(t, z_{\varepsilon_j}(t))dt + B^{\varepsilon_j}(t, z_{\varepsilon_j}(t)) \cdot db^{\varepsilon_j}$$
$$\hat{z}_{\varepsilon_j}(0) = y_0$$

To prove (3.7) amounts to proving that :

$$z_{\varepsilon_j} = \hat{z}_{\varepsilon_j} \qquad \text{a.e.t., a.s.}\omega \tag{3.10}$$

Note also that exactly as for (3.8) we can assert that :

$$E\int_0^T |\hat{z}_{\varepsilon_j}(t) - \hat{z}^{k,\varepsilon_j}(t)|^2 dt \to 0, \qquad \text{as } k \to 0. \tag{3.11}$$

Now we have :

$$E\int_0^T \frac{|\hat{z}_{\varepsilon_j}(t) - z_{\varepsilon_j}(t)|^2}{1 + |\hat{z}_{\varepsilon_j}(t) - z_{\varepsilon_j}(t)|^2} dt \le 2 \;\; E\int_0^T |z_{\varepsilon_j}(t) - \hat{z}_{\varepsilon_j}(t)|^2 dt \tag{3.12}$$
$$+2 \;\; E\int_0^T \frac{|z^{k,\varepsilon_j}(t) - z_{\varepsilon_j}(t)|^2}{1 + |z_{\varepsilon_j}(t) - z_{\varepsilon_j}(t)|^2} dt.$$

But the map

$$b(\cdot), z(\cdot) \to \int_0^T \frac{|z(t) - \phi^{k,\varepsilon}(b(\cdot), z(\cdot))(t)|^2}{1 + |z(t) - \phi^{k,\varepsilon}(b(\cdot), z(\cdot))(t)|^2} dt$$

is continuous and bounded on S. Therefore from (3.4)

$$E\int_0^T \frac{|z^{k,\varepsilon_j}(t) - z_{\varepsilon_j}(t)|^2}{1 + |z_{\varepsilon_j}(t) - z_{\varepsilon_j}(t)|^2} dt = E^{\varepsilon_j}\int_0^T \frac{|y^{k,\varepsilon_j}(t) - y_{\varepsilon_j}(t)|^2}{1 + |y^{k,\varepsilon_j}(t) - y_{\varepsilon_j}(t)|^2} dt$$
$$\le E^{\varepsilon_j}\int_0^T |y^{k,\varepsilon_j}(t) - y_{\varepsilon_j}(t)|^2 dt$$
$$\to 0 \text{ as } k \to 0.$$

This combined with (3.11) yields that the left hand side of (3.12) is 0. Hence (3.10).

From (3.7) and the techniques of Lemmas 2.1, 2.2, 2.3 it is also clear that the following estimates are also valid

$$E\int_0^T \|z_{\varepsilon_j}(t)\|^2 dt \le C \tag{3.13}$$
$$E \sup_{0\le t\le T} |z_{\varepsilon_j}(t)|^2 \le C, \qquad \sup_{0\le t\le T} E|z_{\varepsilon_j}(t)|^4 \le C$$
$$E \sup_{|\theta|\le\delta} \int_0^T \|z_{\varepsilon_j}(t+\theta) - z_{\varepsilon_j}(t)\|_{V^*}^2 dt \le C\delta.$$

Therefore beyond (3.5) we may also assume possibly after extracting a new subsequence, still denoted by ε_j, that

$$(3.14) \qquad \begin{aligned} z_{\varepsilon_j} \to y & \text{ in } L^2(\Omega, \mathcal{A}, \mathcal{P}; L^2(O,T;V)) \text{ weakly} \\ & \text{ in } L^2(\Omega, \mathcal{A}, \mathcal{P}; L^\infty(O,T;H)) \text{ weak star} \\ & \text{ in } L^\infty(0,T; L^4(\Omega, \mathcal{A}, \mathcal{P}; H)) \text{ weak star.} \end{aligned}$$

Furthermore the process y satisfies

$$(3.15) \qquad \begin{aligned} & E\int_0^T \|y(t)\|_V^2 dt \le C, \quad E \sup_{0\le t\le T} |y(t)|^2 \le C, \\ & \sup_{0\le t\le T} E|y(t)|^4 \le C, \\ & E \sup_{|\theta|\le\delta} \int_0^T \|y(t+\theta) - y(t)\|_{V^*}^2 dt \le C\delta. \end{aligned}$$

We also define

$$\mathcal{F}^t = \sigma(\mathbb{1}_t w(\cdot,\omega), \quad \mathbb{1}_t y(\cdot,\omega))$$

and $w(t;\omega)$ is a $\mathcal{F}^t$ Wiener process.

We have also

$$(3.16) \qquad z_{\varepsilon_j} \to y \text{ in } L^2(\Omega, \mathcal{A}, \mathcal{P}; L^2(0,T;H))$$

and thus, by extracting a new subsequence, still denoted in the same way, we can also assert that

$$(3.17) \quad z_{\varepsilon_j} \to y \text{ in for almost all } \omega, t \text{ (with respect to the measure } dP \otimes dt).$$

From the assumptions (1.7), (1.8) it is easy to deduce that

$$(3.18) \qquad g^{\varepsilon_j}(\cdot, z_{\varepsilon_j}(\cdot)) \to g(\cdot, y(\cdot)) \text{ in } L^2(\Omega, \mathcal{A}, \mathcal{P}; L^2(O,T;H))$$

$$(3.19) \qquad B^{\varepsilon_j}(\cdot, z_{\varepsilon_j}(\cdot)) \to B(\cdot, y(\cdot)) \text{ in } L^2(\Omega, \mathcal{A}, \mathcal{P}; L^2(O,T;H^m)).$$

It follows from (3.19) that for any t

$$(3.20) \quad \int_0^t B^{\varepsilon_j}(s, z_{\varepsilon_j}(s)) \cdot db^{\varepsilon_j}(s) \to \int_0^t B(s, y(s)) \cdot dw(s) \text{ in } L^2(\Omega, \mathcal{A}, \mathcal{P}; H) \text{ weakly}$$

(see A. BENSOUSSAN [1]) ; one uses an approximation of $\chi(t) = B(t, y(t))$ as follows

$$\chi^\eta(t) = \frac{1}{\eta}\int_0^t e^{-\frac{t-s}{\eta}} \chi(s) ds.$$

Now

$$(3.21)\int_0^t A^{\varepsilon_j} z_{\varepsilon_j}(s)ds = -z_{\varepsilon_j}(t) + y_0 + \int_0^t g^{\varepsilon_j}(s, z_{\varepsilon_j}(s))ds$$

$$+ \int_0^t B^{\varepsilon_j}(s, z_{\varepsilon_j}(s))db^{\varepsilon_j}$$

$$\to \chi(t) = -y(t) + y_0 + \int_0^t g(s, y(s))ds + \int_0^t B(s, y(s)) \cdot dw(s)$$

in $L^2(\Omega, \mathcal{A}, \mathcal{P}; H)$ weakly.

Let $\xi \in L^2(\Omega, \mathcal{A}, \mathcal{P}; V)$ and ξ^{ε_j} defined by

$$(A^{\varepsilon_j})^* \xi^{\varepsilon_j} = A^* \xi$$

then a.s. $\xi^{\varepsilon_j} \to \xi$ in V weakly, Hstrongly, $L^2(\Omega, \mathcal{A}, \mathcal{P}; V)$ weakly, $L^2(\Omega, \mathcal{A}, \mathcal{P}; H)$ strongly. We can write

$$E < \xi^{\varepsilon_j}, \int_0^t A^{\varepsilon_j} z_{\varepsilon_j}(s)ds > \to E < \chi(t), \xi >$$

and also

$$E < \xi^{\varepsilon_j}, \int_0^t A^{\varepsilon_j} z_{\varepsilon_j}(s)ds > = E \int_0^t < A^* \xi, z_{\varepsilon_j}(s) > ds$$

$$\to E \int_0^t < \xi, Ay(s) > ds.$$

Therefore

$$\chi(t) = \int_0^t Ay(s)ds.$$

Therefore the pair $w(\cdot), y(\cdot)$ satisfies (1.15). The proof of Theorem 1.1 has been completed. □

4. - NONLINEAR FILTERING WITH HOMOGENIZATION

4.1. The problem

We are interested here in Zakaï equation which reads as follows

$$(4.1) \qquad \begin{aligned} &dy^\varepsilon + A^\varepsilon y^\varepsilon dt = B^\varepsilon(t, y^\varepsilon) \cdot dw^\varepsilon \\ &y^\varepsilon(0) = y_0 \end{aligned}$$

where

$$(4.2) \qquad A^\varepsilon = - \text{ div } (a^\varepsilon D\cdot) - \text{ div } (a_0^\varepsilon u)$$

with the assumptions

(4.3) $a^\varepsilon : L^\infty(\mathbb{R}^n; L(\mathbb{R}^n;\mathbb{R}^n))$; $a^\varepsilon\xi\xi \geq \alpha|\xi|^2, (a^\varepsilon)^{-1}\xi\xi \geq \beta|\xi|^2$
with $\alpha, \beta > 0$,

(4.4) $a_0^\varepsilon(x) : L^\infty(\mathbb{R}^n;\mathbb{R}^n)$ $\|a_0^\varepsilon\| \leq c_0$

(4.5) $B^\varepsilon(t,u)(x) = u(x)h^\varepsilon(x,t)$
with $h^\varepsilon : \mathbb{R}^n \times (0,T) \to \mathbb{R}^m$, $\|h^\varepsilon\| \leq c_1$.

Define

$$g_i^\varepsilon = \frac{\partial a_{ij}^\varepsilon}{\partial x_j} - a_{0,i}^\varepsilon, \qquad \sigma^\varepsilon = (2a^\varepsilon)^{1/2} \tag{4.6}$$

Equation (4.1) arises in the context of the following nonlinear filtering problem.

Consider the diffusion

$$\begin{aligned} dx^\varepsilon &= g^\varepsilon(x^\varepsilon)dt + \sigma^\varepsilon(x^\varepsilon)\cdot d\beta^\varepsilon \\ x^\varepsilon(0) &= \xi \end{aligned} \tag{4.7}$$

and the observation process

$$\begin{aligned} dw^\varepsilon &= h^\varepsilon(x^\varepsilon,t)dt + d\gamma^\varepsilon \\ w^\varepsilon(0) &= 0 \end{aligned} \tag{4.8}$$

where $\beta^\varepsilon, w^\varepsilon$ are independent Wiener processes with values in $\mathbb{R}^n$ and $\mathbb{R}^m$ respectively, defined on a convenient system $\Omega^\varepsilon, \mathcal{A}^\varepsilon, \mathcal{P}^\varepsilon, \mathcal{F}^{t,\varepsilon}$.

We consider that ξ is $\mathcal{F}^{0,\varepsilon}$ measurable and has a probability density with respect to Lebesgue measure, which is $y_0 \in L^2(\mathbb{R}^n)$. Define $H = L^2(\mathbb{R}^n)$, $V = H^1(\mathbb{R}^n)$, the solution of (4.1) defined in $L^2_{\mathcal{F}^{t,\varepsilon}}(0,T;V) \cap L^2(\Omega^\varepsilon, \mathcal{A}^\varepsilon, \mathcal{P}^\varepsilon, C(0,T;H))$ coincides with the unconditional probability density related to the problem (4.6), (4.7). We shall need Sobolev spaces with weights. Let

$$\pi_s(x) = (1+|x|^2)^{s/2}$$

and set

$$\begin{aligned} H_0 &= \{u | u\pi_s \in H\} \\ H_0^* &= \{u | u\pi_s^{-1} \in H\}. \end{aligned}$$

Clearly one has

$$H_0 \subset H \subset H_0^*$$

algebraically and topologically, each space being dense in the next one. Similarly let

$$\begin{aligned} V_0 &= \{u | \pi_s u \in V\} \\ V_0^* &= \{u | \pi_s^{-1} u \in V^*\} \end{aligned}$$

provided with the norms

$$\|u\|_0 = \Big(\int (\pi_s u)^2 dx + \int \pi_s^2 |Du|^2 dx\Big)^{1/2}$$

and its dual, which is expressed as follows. If $\psi \in V_0^*$, then

$$< \psi, u >= (\psi_0, u) + \sum_i (\psi_i, \frac{\partial u}{\partial x_i})$$

and $\psi_0, \psi_1 \dots \psi_n \in H_0^*$

$$\|\psi\|_{V_0^*} = \Big(\sum_{i=0}^{n} |\psi_i|_{H_0^*}^2\Big)^{1/2}.$$

We have the following inclusions

$$\begin{aligned} &V_0 \subset H_0 \subset H \subset H_0^* \subset V_0^* \\ &V_0 \subset V \subset H \subset V^* \subset V_0^* \end{aligned} \tag{4.9}$$

algebraically and topologically, each space being dense in the next one.

The interesting compactness property is the following

(4.10) the injection of V in H_0^* is compact.

We shall indicate by $\|\ \|_0$, $\|\ \|_{*,0}$ and $|\ |_0$, $|\ |_{*,0}$ the norms in V_0, V_0^*, H_0, H_0^*.

We shall replace the assumption (1.2) by the following. Let

$$\begin{aligned} W_0 &= \{u | \pi_s^2 u \in V\} \\ W_0^* &= \{u | \pi_s^{-2} u \in V^*\} \\ \|u\|_{W_0}^2 &= \int \pi_s^4 u^2 dx + \int \pi_s^4 |Du|^2 dx \end{aligned}$$

and the injection of W_0 in H_0 is compact.

There exists

(4.11) $\quad a : L^\infty(\mathbb{R}^n \times (0,T); L(\mathbb{R}^n;\mathbb{R}^n)), \quad a_0 : L^\infty(\mathbb{R}^n \times (0,T); \mathbb{R}^n)$

with $a\xi \cdot \xi \geq \alpha_1 |\xi|^2$, $\alpha_1 > 0$, such that if γ is large enough, $\forall u \in W_0$, the solution of

$$(A^\varepsilon)^* u_\varepsilon + \gamma u_\varepsilon = A^* u + \gamma u$$

verifies $u_\varepsilon \in W_0$ and $u_\varepsilon \to u$ in W_0 weakly where we have set

$$\begin{aligned} (A^\varepsilon)^* &= -\operatorname{div}\,((a^\varepsilon)^* D\cdot) + a_0^\varepsilon \cdot D \\ A^* &= -\operatorname{div}\,(a^* D\cdot) + a_0 \cdot D. \end{aligned}$$

Moreover one has

$$h^\varepsilon \to h \text{ in } L^\infty(\mathbb{R}^n \times (0,T);\mathbb{R}^n) \text{ pointwise.} \tag{4.12}$$

We set

$$B(t,u)(x) = u(x)h(x,t). \tag{4.13}$$

4.2. The main result

Consider the equation

$$\begin{aligned} dy + Ay\ dt &= B(t,y)\cdot dw \\ y(0) &= y_0. \end{aligned} \tag{4.14}$$

A solution of (4.14) is a system $(\Omega, \mathcal{A}, \mathcal{P}, \mathcal{F}^t, w(t), y(t))$ such that w is a $\mathcal{F}^t$ standard Wiener process, $y \in L^2_{\mathcal{F}}(0,T;V) \cap L^2(\Omega,\mathcal{A},\mathcal{P};C(0,T;H))$, and (4.13) is satisfied.

Note that it is well known (see E. PARDOUX [7] or A. BENSOUSSAN [2]) that to study (4.14) we can take $\Omega, \mathcal{A}, \mathcal{P}, \mathcal{F}^t, w(\cdot)$ arbitrary and the corresponding $y(\cdot)$ is uniquely defined.

We set $S = C(0,T;\mathbb{R}^m) \times L^2(0,T;H_0^*)$, and call π^ε, π the probability measures on S associated to $(w^\varepsilon(\cdot), y^\varepsilon(\cdot))$ and $(w(\cdot),y(\cdot))$ solutions of (4.1), (4.14), respectively. Since y^ε and y are uniquely defined, the probability measures π^ε and π are also uniquely defined.

We can assert

Theorem 4.1. *Assume (4.3), (4.4), (4.5), (4.11), (4.12), and $y_0 \in L^2$. Then $\pi^\varepsilon \to \pi$ in the sense of weak topology of probability measures on S.*

5. - PROOF OF THEOREM 4.1

5.1. A priori estimates

We have the

Lemma 5.1. *The following estimates hold*

$$E^\varepsilon \int_0^T \|y^\varepsilon(t)\|^2 dt \le C \tag{5.1}$$

$$E^\varepsilon |y^\varepsilon(t)|^2 \le C, \quad E^\varepsilon |y^\varepsilon(t)|^4 \le C. \tag{5.2}$$

Proof.

We make use of energy equality

$$E^\varepsilon |y^\varepsilon(t)|^2 + 2E^\varepsilon \int_0^t < A^\varepsilon y^\varepsilon, u^\varepsilon > ds = |y_0|^2 + E^\varepsilon \int_0^t |B^\varepsilon(y^\varepsilon)|^2 ds \tag{5.3}$$

and proceed as in Lemma 2.1, making use of Gronwall's inequality. Details are omitted. □

Lemma 5.2. *We have the estimate*

$$E^\varepsilon \sup_{t\in[0,T]} |y^\varepsilon(t)|^2 \leq C. \tag{5.4}$$

Proof

Analogous to that of Lemma 2.2. □

Lemma 5.3. *We have the estimate*

$$E^\varepsilon \sup_{|\theta|\leq\delta} \int_0^T \|y^\varepsilon(t+\theta) - y^\varepsilon(t)\|^2_{V^*} dt \leq C\delta, \quad \delta < 1 \tag{5.5}$$

Proof

Same as in Lemma 2.3. □

5.2. Tightness property

In the Hilbert space $L^2(0,T;H_0^*)$ we consider a set Z as in (3.1), namely

$$\begin{aligned} Z = \{z| \int_0^T \|z(t)\|^2_V dt \leq K, \quad |z(t)|^2_H \leq L \text{ a.e.t.}, \\ \sup_{|\theta|\leq\mu_n} \int_0^T \|z(t+\theta) - z(t)\|^2_{V_0^*} dt \leq \nu_n M, \quad \forall n\} \end{aligned} \tag{5.6}$$

where $\mu_n, \nu_n \geq 0$ and $\mu_n, \nu_n \to 0$.

In the definition of Z, z is extended by 0 outside $(0,T)$. The set Z is a compact subset of $L^2(0,T;H_0^*)$.

Considering $S = C(0,T;R^m) \times L^2(0,T;H_0^*)$, we define on S the probability measure π^ε image on S of the probability P^ε on Ω^ε, by the map

$$\omega \in \Omega^\varepsilon \Rightarrow w_\varepsilon(\cdot), y_\varepsilon(\cdot)$$

and we have the

Lemma 5.4. *The family π^ε is uniformly tight.*

Proof.

Similar to that of Lemma 3.1. □

5.3. Convergence

We extract a subsequence such that

$$\pi^{\varepsilon j} \to \pi, \quad \text{i.e.} \forall \phi : S \to \mathbb{R}, \tag{5.7}$$

continuous and bounded,

$$\int \phi(b(\cdot), z(\cdot)) d\pi^{\varepsilon j} \to \int \phi(b(\cdot), z(\cdot)) d\pi.$$

We construct also a probability space $(\Omega, \mathcal{A}, \mathcal{P})$ and random variables, $w(\cdot, \omega)$, $y(\cdot, \omega)$, $b_{\varepsilon_j}(\cdot, \omega)$, $z_{\varepsilon_j}(\cdot, \omega)$, on $(\Omega, \mathcal{A}, \mathcal{P})$ with values in S such that

(5.8) the probability law of $b_{\varepsilon_j}(\cdot), z_{\varepsilon_j}(\cdot)$ is π^{ε_j}

(5.9) $b_{\varepsilon_j}(\cdot, \omega), z_{\varepsilon_j}(\cdot, \omega) \to w(\cdot, \omega), y(\cdot, \omega)$ a.s. in S

(5.10) the probability law of $w(\cdot, \omega), y(\cdot, \omega)$ is π.

We can again state that z_{ε_j} is the solution of

$$\begin{aligned} & dz_{\varepsilon_j} + A^{\varepsilon_j} z_{\varepsilon_j} dt = B^{\varepsilon_j}(t, z_{\varepsilon_j}) \cdot db_{\varepsilon_j} \\ & z_{\varepsilon_j}(0) = y_0. \end{aligned} \tag{5.11}$$

We have moreover the estimates

$$\begin{aligned} & E \int_0^T \|z_{\varepsilon_j}(t)\|^2 dt \leq C \\ & E \sup_{0 \leq t \leq T} |z_{\varepsilon_j}(t)|^2 \leq C, \quad \sup_{0 \leq t \leq T} E|z_{\varepsilon_j}(t)|^4 \leq C \\ & E \sup_{|\theta| \leq \delta} \int_0^T \|z_{\varepsilon_j}(t+\theta) - z_{\varepsilon_j}(t)\|_{V^*}^2 dt \leq C\delta. \end{aligned} \tag{5.12}$$

Therefore we may also assume, possibly after extracting a new subsequence, still denoted by ε_j, that

$$\begin{aligned} z_{\varepsilon_j} \to y \text{ in } & L^2(\Omega, \mathcal{A}, \mathcal{P}; L^2(0,T;V)) \text{ weakly} \\ \text{in } & L^2(\Omega, \mathcal{A}, \mathcal{P}; L^\infty(0,T;V)) \text{ weak star} \\ \text{in } & L^\infty(0,T; L^4(\Omega, \mathcal{A}, \mathcal{P}; H)) \text{ weak star} \end{aligned} \tag{5.13}$$

and also

$$\begin{aligned} & z_{\varepsilon_j} \to y \text{ in } L^2(\Omega, \mathcal{A}, \mathcal{P}; L^2(0,T;H_0^*)) \\ & \text{and } z_{\varepsilon_j}(t,\omega) \to y(t,\omega) \text{ in } H_0^* \end{aligned} \tag{5.14}$$

for almost all ω, t.

The process y satisfies

$$E\int_0^T \|y(t)\|_V^2 dt \leq C \tag{5.15}$$
$$E \sup_{0\leq t\leq T} |y(t)|^2 \leq C, \qquad \sup_{0\leq t\leq T} E|y(t)|^4 \leq C,$$
$$E \sup_{|\theta|\leq\delta} \int_0^T \|y(t+\theta)-y(t)\|_{V^*}^2 dt \leq C\delta.$$

We have also

$$B^{\varepsilon_j}(t, z_{\varepsilon_j}) \to B(t,y) \text{ in } L^2(\Omega, \mathcal{A}, \mathcal{P}; L^2(0,T;(H_0^*)^m)). \tag{5.16}$$

Therefore

$$\forall t, \int_0^t B^{\varepsilon_j}(s, z_{\varepsilon_j}(s))db_{\varepsilon_j} \to \int_0^t B(s,y(s))\cdot dw \text{ in } L^2(\Omega,\mathcal{A},\mathcal{P};H_0^*) \text{ weakly.} \tag{5.17}$$

Let next $\zeta \in L^2(\Omega,\mathcal{A},\mathcal{P};W_0)$ and ζ_ε defined by

$$(A^\varepsilon)^*_{\zeta_\varepsilon} + \gamma\zeta_\varepsilon = A^*\zeta + \gamma\zeta$$

as in (4.15). Since

$$\|\zeta_\varepsilon\|_{W_0} \leq C\|\zeta\|_{W_0}$$

it follows that a.s. $\zeta_\varepsilon \to \zeta$ in W_0 weakly and H_0 strongly, and in $L^2(\Omega,\mathcal{A},\mathcal{P};H_0)$.

We write from (5.11)

$$(\zeta_{\varepsilon_j}, z_{\varepsilon_j}(t)) + \int_0^t < (A^{\varepsilon_j})^*\zeta_{\varepsilon_j}, z_{\varepsilon_j} > ds = (\zeta_{\varepsilon_j}, \int_0^t B^{\varepsilon_j}(s, z_{\varepsilon_j})db_{\varepsilon_j})$$

or

$$E(\zeta_{\varepsilon_j}, z_{\varepsilon_j}(t)) + E\int_0^t < A^*\zeta, z_{\varepsilon_j} > ds$$
$$= \gamma E\int_0^t (\zeta-\zeta_\varepsilon, z_{\varepsilon_j})ds = E(\zeta_{\varepsilon_j}, \int_0^t B^{\varepsilon_j}(s, z_{\varepsilon_j})db_{\varepsilon_j}).$$

For almost all t, we can pass to the limit, to obtain

$$E(\zeta, y(t)) + E\int_0^t < \zeta, Ay > ds = E(\zeta, \int_0^t B(s,y(s))\cdot dw(s)).$$

Modifying $y(t)$ in a set of Lebesgue measure 0, we deduce easily that $y(t), w(t)$ is a solution of (4.14). This completes the proof of Theorem 4.1., since any subsequence of π^ε will converge to π. □

5.4. Additional remarks

Classical homogenization

Assume $a^\varepsilon(x) = a(\frac{x}{\varepsilon})$ with $a(y)$ periodic with period 1 in each component, and $a_0^\varepsilon(x) = 0$.

Consider

$$-\frac{\partial}{\partial x_i}(a_{ji}(\frac{x}{\varepsilon})\frac{\partial u_\varepsilon}{\partial x_j}) + \gamma u_\varepsilon = f \tag{5.18}$$

with $f\pi_s^\varepsilon \in V^*$. Note that $| < f, v\pi_s^4 > | \leq C\|f\pi_s^2\|_{V^*}\|v\|_{W_0}$ where if $f \in V^*$, $f = (f_0, f_1, \ldots, f_n)$ we define $f\pi_s^2 = (f_0\pi_s^2, f_1\pi_s^2, \ldots, f_n\pi_s^2)$, and $f\pi_s^2 \in V^*$ if $f_i\pi_s^2 \in H, \forall i = 0 \ldots n$.

Multiplying (5.18) by $\pi_s^4 u_\varepsilon$ we deduce

$$\int a_{ji}(\frac{x}{\varepsilon})\frac{\partial u_\varepsilon}{\partial x_j}\frac{\partial u_\varepsilon}{\partial x_i}\pi_s^4 dx + \int a_{ji}(\frac{x}{\varepsilon})\frac{\partial u_\varepsilon}{\partial x_j}u_\varepsilon\pi_s^4\frac{x_i}{1+|x|^2}dx$$
$$+\gamma\int u_\varepsilon^2\pi_s^4 dx \leq C\|f\pi_s^2\|_{V^*}\|u_\varepsilon\|_{W_0}$$

hence

$$\|u_\varepsilon\|_{W_0} \leq C$$

if γ is chosen large enough.

The assumption (4.11) is then satisfied by classical homogenization theory (see B.L.P. [4]).

Remark on h^ε

The pointwise convergence (4.12) is rather restrictive. It forbids an oscillatory behavior similar to those of a^ε and a_0^ε. In this case we proved a weaker result in BENSOUSSAN-BLANKENSHIP [3]. Consider the model (4.1) with a fixed Wiener process $w^\varepsilon = w$, i.e.

$$dy^\varepsilon + A^\varepsilon y^\varepsilon dt = B^\varepsilon(t, y^\varepsilon)\cdot dw$$
$$y^\varepsilon(0) = y_0$$

Assume that $\mathcal{F}^t = \sigma(w(s), s \leq t)$. Let $\psi(x)$ smooth bounded on $\mathbb{R}^n$ and $\beta(t)$ smooth bounded function on $[0,T]$ with values in $\mathbb{R}^n$. Consider the parabolic equation

$$\frac{\partial V^\varepsilon}{\partial t} + \text{div}\,(a^{\varepsilon *}DV^\varepsilon) - a_0^\varepsilon\cdot DV^\varepsilon + V^\varepsilon h^\varepsilon\cdot\beta = 0$$
$$V^\varepsilon(x,T) = \psi(x)$$

Suppose that $V^\varepsilon(x,t) \to V(x,t)$ pointwise, where V is the solution of

$$\frac{\partial V}{\partial t} + \text{div}\,(aDV) - a_0 DV + Vh\cdot\beta = 0$$

$$V(x,T) = \psi(x)$$

then $\int y^\varepsilon(x,T)\psi(x)dx \to \int y(x,T)\psi(x)dx$ in $L^2(\Omega,\mathcal{A},\mathcal{P})$ weakly, where y is the solution of

$$dy + Ay\ dt = B(t,y)\cdot dw$$

$$y(0) = y_0.$$

This is a much weakier result than that of Theorem 4.1. It means roughly speaking that

$$E\phi(y^\varepsilon(\cdot),w(\cdot)) \to E\phi(y(\cdot),w(\cdot)) \tag{5.19}$$

for functionals $\phi(z(\cdot),w(\cdot))$ of the form

$$\phi(z(\cdot),w(\cdot)) = \int_0^T (z(t),\varphi(t))dt \cdot \chi(x(\cdot))$$

where χ continuous bounded on $C(0,T;\mathbb{R}^m)$, $\varphi \in L^\infty(0,T;H_0)$ whereas Theorem 4.1 means the convergence (5.19) for all continuous bounded functionals.

References

[1] A. BENSOUSSAN, Some Existence Results for Stochastic Partial Differential Equations, to be published.

[2] A. BENSOUSSAN, *Stochastic Control with partial observation*, Cambridge University Press, Cambridge, 1990.

[3] A. BENSOUSSAN, G. BLANKENSHIP, Nonlinear Filtering with Homogenization, Stochastics, 1986, vol. 17, pp. 67-90.

[4] A. BENSOUSSAN, J.L. LIONS, G. PAPANICOLAOU, *Asymptotic Analysis for Periodic Structures*, North-Holland, Amsterdam, 1978.

[5] E. DE GIORGI, S. SPAGNOLO, Sulla convergenza degli integrali dell' energia per operatori ellitici del secondo ordine, Boll. U.M.I. 8 (1973), 391-411.

[6] N. NAGASE, On the Cauchy problem for nonlinear stochastic partial differential equations with continuous coefficients, Existence theorems, to be published.

[7] E. PARDOUX, Stochastic partial differential equations and filtering of diffusion processes, Stochastics, 3 (1979), pp. 127-167.

[8] Y.V. PROKHOROV, Convergence of random processes and limit theorems in probability theory, Theory Prob. Appl. 1 (1956), 157-214.

[9] A.V. SKOROKHOD, *Studies in the Theory of Random Processes*, Addison Wesley, Reading, Mass. 1965.

[10] L. TARTAR, Cours Peccot, Collège de France, 1977.

A. Bensoussan, University of Paris Dauphine and INRIA

EFFECTIVE MEDIUM APPROXIMATION FOR NONLINEAR CONDUCTIVITY OF A COMPOSITE MEDIUM

David J. Bergman

Abstract

Bruggeman's Self-Consistent Effective Medium Approximation for the linear conductivity of a composite medium is reformulated in a manner such that the averaging procedure that must be used becomes unambiguous. The new formulation can be applied equally unambiguously to nonlinear conductivity in a composite medium. It is used to construct a Bruggeman-type approximation for a strong, power law nonlinear conductivity, and the scaling form of this approximation in the vicinity of a percolation threshold is found. It is also used to construct such an approximation for the small power-law-nonlinearity correction term in a composite conductor whose leading behavior is linear. In this case the Bruggeman-type approximation becomes useless long before the percolation threshold is reached. The reasons for this are discussed.

1. Introduction

The Self-Consistent Effective Medium Approximation was invented by Bruggeman in 1935 [1], and has remained one of the most popular and ubiquitous approximations used for calculating the usual (i.e., the linear) bulk effective electrical conductivity σ_e of a two-component composite medium. This is due, in large measure, to the simplicity of the approximation, to its intuitive appeal, and to the fact that it gives accurate results when either component is very sparse. It also has a non-trivial percolation threshold, which most other simple approximations do not possess. Finally, none of the complicated details of the microstructure are used in constructing this approximation. The only information that is needed are the values of the component conductivities σ_1,σ_2, the component volume fractions p_1,p_2, the fact that the individual grains are spherical, and the fact that the composite is macroscopically homgeneous.

It is trivial to generalize this approximation to a multicomponent composite, or to a medium in which the local conductivity $\sigma(\vec{r})$ is a continuous function of position - the main price is that the equation which must be solved to determine σ_e is no longer a simple quadratic equation. It is also quite straightforward to generalize the approximation to encompass ellipsoidal grains with either isotropic or anisotropic spatial distributions, or to the case when the component conductivities σ_i are non-scalar tensors - in both cases σ_e is also a non-scalar tensor [2]. Even the case when the σ_i, and therefore also σ_e, are non-symmetric tensors can be treated, which is the case when an external magnetic field is present [3].

Bruggeman's approximation applies without any changes also to the dielectric susceptibility, magnetic permeability, thermal conductivity, and chemical diffusivity coefficients, since the mathematical structure of the associated physical properties is the same as that of electrical conduction. It is also easily extended to the case when any of these moduli is complex. With simple modifications it can be applied to discrete network models of these phenomena [4]. Effective medium approximations of the same type have been formulated for bulk effective moduli that characterize other types of linear response of a composite medium. These include elastic stiffness moduli [5,6] and thermoelectric coefficients [7].

In view of all this, it is quite natural to attempt to generalize Bruggeman's approximation to the domain of nonlinear properties of composites. In order to do this we review and then reformulate this approximation in a manner that can be applied unambiguously to nonlinear behavior (Section 2). We then apply this approach to a strong, power law, nonlinear conductivity (Section 3). This leads to results which are quite satisfactory including a scaling theory near a percolation threshold. We then attempt to apply the same approach to the case where the electrical conductivity is linear in leading order, but has a small nonlinear power law correction term (a so called weak nonlinearity) (Section 4). Although a very definite and unambiguous result is obtained, this result sometimes diverges. It turns out that in this case one sometimes cannot get a useful result without including some information about details of the microstructure.

2. Bruggeman's Approximation - Review and Reformulation

In order to maximize the intuitive clarity of all the considerations, we present the approximation in the context of dielectric behavior. The composite dielectric is replaced by a fictitious homogeneous dielectric with a dielectric constant equal to the (as yet unknown) bulk effective dielectric constant of the composite ϵ_e. If a uniform electric field $\vec{E}_0$ is applied to this fictitious material, and a spherical inclusion made of one of the component dielectrics ϵ_i is inserted, it acquires an induced dipole moment given by

$$\text{(2.1)} \qquad \vec{p} = a^3 \frac{\epsilon_i - \epsilon_e}{\epsilon_i + 2\epsilon_e} \vec{E}_0 \quad ,$$

where a is the radius of the inclusion. The value of ϵ_e is then determined by imposing the consistency requirement that the *arithmetic* average of $\vec{p}$ over the different possible types of inclusion must vanish (p_i is the volume fraction of the i - component)

(2.2) $$0 = \sum_i p_i \frac{\epsilon_i - \epsilon_e}{\epsilon_i + 2\epsilon_e} .$$

For a two component composite this leads to a quadratic equation for ϵ_e, for which the correct solution is the one that is positive when ϵ_1,ϵ_2 are also positive

$$\epsilon_e = \frac{3}{4}\left\{ \epsilon_1\left(p_1 - \frac{1}{3}\right) + \epsilon_2\left(p_2 - \frac{1}{3}\right) + \right.$$

$$\left. + \left[\epsilon_1^2\left(p_1 - \frac{1}{3}\right)^2 + \epsilon_2^2\left(p_2 - \frac{1}{3}\right)^2 + 2\epsilon_1\epsilon_2\left(\frac{2}{9} + p_1p_2\right)\right]^{1/2} \right\} .$$

(2.3)

If we take $\epsilon_2 = 0<\epsilon_1$, this reduces to

(2.4) $$\epsilon_e = \begin{cases} \frac{3}{2}\epsilon_1(p_1 - \frac{1}{3}) & p_1 > \frac{1}{3} \\ 0 & p_1 < \frac{1}{3} \end{cases} ,$$

which clearly exhibits the effective medium percolation threshold at $p_1 = \frac{1}{3}$ (it is intiuitvely useful to think of the last equation as describing conductivity rather than dielectric constant).

If we had used a different type of average when implementing the consistency requirement, we would have got a different result. We therefore examine the reasons for this choice. It seems that the only reason for preferring the arithmetic average is that for a dilute system of inclusions, the effects of different inclusions (e.g., on the local electric field $\vec{E}(\vec{r})$) are approximately additive. In view of the fact that we apply this procedure to non-dilute systems, this appears to be a somewhat unconvincing argument. It appears even less convincing when we turn to consider the case of a nonlinear dielectric.

As an alternative to the line of reasoning presented above, we start by considering, as before, a fictitious homogeneous dielectric, whose dielectric constant is now denoted by ϵ_0. When a single spherical inclusion of volume $V_a = 4\pi a^3/3$ and dielectric constant ϵ_i is inserted to form a composite medium, the bulk effective dielectric constant can be calculated from

$$(2.5) \qquad \epsilon_e = \frac{1}{V} \int dV \, \epsilon(\vec{r}) \left[\frac{\vec{E}(\vec{r})}{E_0} \right]^2 ,$$

where V is the total volume and is much greater than V_a, $\epsilon(\vec{r})$ is the position dependent dielectric constant ($\epsilon(\vec{r})$ is either ϵ_0 or ϵ_i), and $\vec{E}(\vec{r})$ is the local electric field when the volume average (or externally applied) field is $\vec{E}_0$. For finite V it is difficult to calculate (2.5) exactly, and the result would depend on the shape of the sample and on the precise positioning of the inclusion. But up to and including terms of order V_a/V, the result is independent of those details, and one gets, by an elementary calculation [8],

$$(2.6) \qquad \epsilon_e = \epsilon_0 + 3\epsilon_0 \frac{V_a}{V} \frac{\epsilon_i - \epsilon_0}{\epsilon_i + 2\epsilon_0} + \mathcal{O}\left(\frac{V_a^2}{V^2} \right) .$$

If one now imposes the consistency requirement that the average of ϵ_e over the different types of inclusion must be equal to ϵ_0, one immediately recovers (2.2) and hence (2.3). But because the leading term in (2.6) is a constant, and it is only the small $\mathcal{O}(V_a/V)$ correction which fluctuates from sample to sample, it does not matter what kind of average is being used. Therefore this procedure can immediately be extended to the case of a composite dielectric or conducting medium that is nonlinear [8].

3. Composite Conductor with a Strong Power Law Nonlinearity

We consider a composite conducting medium where each component exhibits the following isotropic power law relation between the electric field $\vec{E}(\vec{r})$ and current density $\vec{J}(\vec{r})$

$$(3.1) \qquad \vec{J} = \sigma |E|^\beta \vec{E} .$$

Here the nonlinearity exponent $\beta \neq 0$ is assumed to have the same value in all components, but the nonlinear conductivity σ is different. By considering the local rate of production of Joule heat $\vec{E} \cdot \vec{J}$ integrated over the entire sample, and equating it to the rate produced by a fictitious homogenous medium of the same total size and shape, one easily finds that in order to produce the same amount of heat, the homogeneous medium must have a conductivity given by

$$(3.2) \qquad \sigma_e = \frac{1}{V} \int dV\, \sigma(\vec{r}) \left| \frac{\vec{E}(\vec{r})}{E_0} \right|^{\beta+2} .$$

Eq. (2.5) is clearly a special case of this equation. We note that this expression is variational with respect to the field $\vec{E}(\vec{r})$. The determination of $\vec{E}(\vec{r})$ necessitates numerical solution of a nonlinear partial differential equation for the scalar potential $\phi(\vec{r})$ (since $\vec{E}(\vec{r}) = -\nabla\phi(\vec{r})$), namely,

$$(3.3) \qquad 0 = \nabla \cdot \vec{J} \;= \nabla \cdot (\sigma(\vec{r}) \left|\nabla\phi\right|^{\beta} \nabla\phi) .$$

Even without knowing the exact form of ϕ, it is obvious from (3.3) that $\phi(\vec{r}, \sigma_1, \sigma_2 ...)$ is a homogeneous function of order 0 of the component conductivities σ_i. From this and from (3.2) it then follows that $\sigma_e(\sigma_1, \sigma_2 ...)$ is a homogeneous function of order 1 for a composite medium with a given microgeometry.

In order to derive a Bruggeman-type approximation for σ_e, we again consider a large volume V of the fictitious homogeneous conductor σ_0 with a single spherical inclusion of the σ_i component of volume $V_a \ll V$. In contrast with the linear case discussed in the previous section, (3.3) cannot be solved analytically for this case even when we restrict ourselves to $\mathcal{O}(V_a/V)$ terms. From the homogeneity properties of $\sigma_e(\sigma_0,\sigma_i)$ we can however conclude that σ_e has the following form

$$(3.4) \qquad \sigma_e = \sigma_0 + \sigma_0 \frac{V_a}{V} f\left(\frac{\sigma_0}{\sigma_i}\right) + \mathcal{O}\left(\frac{V_a^2}{V^2}\right) ,$$

where f(x) is a function that remains to be determined. Imposing the consistency condition, that the average of σ_e over the various types of inclusions be equal to σ_0, leads to the following equation

$$(3.5) \qquad p_1 f\left(\frac{\sigma_e}{\sigma_1}\right) + (1 - p_1) f\left(\frac{\sigma_e}{\sigma_2}\right) = 0$$

in the case of a two-component composite. Here we have replaced σ_0 by σ_e. Eq. (3.5) implicitly defines the Effective Medium or Bruggeman-Type Approximation for the nonlinear effective conductivity σ_e.

In order to identify the percolation threshold p_{1c} in this approximation, we must consider the case when the constrast between the components is large i.e., $\sigma_2 \ll \sigma_1$. Near the threshold we expect to find

$$\sigma_2 \ll \sigma_e \ll \sigma_1 \ . \tag{3.6}$$

Solving (3.5) for p_1 and taking the limits

$$\frac{\sigma_2}{\sigma_e} \to 0 \ , \ \frac{\sigma_e}{\sigma_1} \to 0 \ , \tag{3.7}$$

we find

$$p_{1c} = \frac{f(\infty)}{f(\infty) - f(0)} \ . \tag{3.8}$$

In order to determine the form of the critical behavior of σ_e near p_{1c}, we need to determine the forms of f(x) both for x near 0 and x near ∞.

For small x, we start by observing that when $\sigma_0 \ll \sigma_1$, a qualitative consideration based on the continuity of the normal component of $\vec{J}(\vec{r})$ at the surface of the inclusion leads to the result that the magnitude of the field $\vec{E}(\vec{r})$ satisfies

$$\left|\frac{E}{E_0}\right|^{\beta+1} \sim \begin{cases} \dfrac{\sigma_0}{\sigma_1} & \text{inside the } \sigma_1 \text{ inclusion} \\ 1 & \text{inside the } \sigma_0 \text{ host} \end{cases} \ . \tag{3.9}$$

Expressing $\sigma(\vec{r})$ with the help of the characteristic function of the host medium $\theta_0(\vec{r})$ and that of the inclusion $\theta_1(\vec{r})$

$$\sigma(\vec{r}) = \sigma_0\theta_0 + \sigma_1\theta_1 \ , \tag{3.10}$$

we can rewrite (3.2) as

$$\frac{\sigma_e}{\sigma_0} = \frac{1}{V}\int dV \left(\theta_0 + \frac{\sigma_1}{\sigma_0}\theta_1\right)\left|\frac{E}{E_0}\right|^{\beta+2} \ , \tag{3.11}$$

which is a function only of σ_0/σ_1. Using (3.9) to evaluate the two terms in (3.11), and comparing the result with (3.4), we get, for x<<1,

$$f(x) \simeq f(0) - a\,(\beta+1)\, x^{\frac{1}{\beta+1}}$$

$$f(0) = \frac{1}{V_a} \int dV\, \theta_0 \left(\left| \frac{E}{E_0} \right|^{\beta+2} - 1 \right) - 1$$

$$a = \lim_{\frac{\sigma_0}{\sigma_1} \to 0} \frac{1}{V_a} \int dV\, \theta_1 \left| \frac{E}{E_0} \right|^{\beta+2} \left(\frac{\sigma_1}{\sigma_0}\right)^{\frac{\beta+2}{\beta+1}} .$$

(3.12)

Both f(0) and a are $\mathcal{O}(1)$, and $f(0)>0$ since $\sigma_1>\sigma_0$ entails $\sigma_e>\sigma_0$.

For large x we begin by noting that when $\sigma_0>>\sigma_2$, qualitative reasoning based on the continuity of the tangential component of $\vec{E}(\vec{r})$ at the surface of the inclusion leads to the following result for the magnitude of $\vec{E}(\vec{r})$

(3.13) $$\left| \frac{E}{E_0} \right| \sim 1 \text{ everywhere} .$$

Proceeding in analogy with (3.10)-(3.12) we write

$$\sigma(\vec{r}) = \sigma_0\theta_0 + \sigma_2\theta_2$$

$$\frac{\sigma_e}{\sigma_0} = \frac{1}{V} \int dV \left(\theta_0 + \frac{\sigma_2}{\sigma_0}\,\theta_1\right) \left| \frac{E}{E_0} \right|^{\beta+2} ,$$

(3.14)

and finally we get, for $x>>1$,

$$f(x) \simeq f(\infty) + \frac{b}{x}$$

$$f(\infty) = \frac{1}{V_a}\int dV\,\theta_0\left(\left|\frac{E}{E_0}\right|^{\beta+2} - 1\right) - 1$$

$$b = \frac{1}{V_a}\int dV\theta_2\left|\frac{E}{E_0}\right|^{\beta+2} .$$

(3.15)

As before, both $f(\infty)$ and b are $\mathcal{O}(1)$, while $f(\infty)<0$ since $\sigma_2<\sigma_0$ entails $\sigma_e<\sigma_0$.

Using thse results in (3.5), together with (3.8), we get

$$(3.16)\quad p_1-p_{1c} \simeq \left(1 + \frac{f(0)-a(1+\beta)\left(\frac{\sigma_e}{\sigma_1}\right)^{\frac{1}{1+\beta}}}{-f(\infty) - b\,\frac{\sigma_2}{\sigma_e}}\right)^{-1} - \left(1 - \frac{f(0)}{f(\infty)}\right)^{-1} ,$$

where σ_0 has again been replaced by σ_e. Assuming that both $\sigma_e/\sigma_1<<1$ and $\sigma_2/\sigma_e<<1$, while $f(0)$, $|f(\infty)|$, a,b,β are all $\mathcal{O}(1)$, we can approximate this equation by

$$(3.17)\quad p_1-p_{1c} \simeq \left(1 + \frac{f(0)}{|f(\infty)|}\right)^2\left[\frac{a(1+\beta)}{f(0)}\left(\frac{\sigma_e}{\sigma_1}\right)^{\frac{1}{1+\beta}} - \frac{b}{|f(\infty)|}\,\frac{\sigma_2}{\sigma_e}\right] .$$

This can be rewritten as

$$1 \simeq p_{1c}{}^2\left(\frac{\sigma_e/\sigma_1}{\Delta p^{1+\beta}}\right)^{\frac{1}{1+\beta}}\left|\frac{a(1+\beta)}{f(0)} - \frac{b}{|f(\infty)|}\left(\frac{\sigma_e/\sigma_1}{\Delta p^{1+\beta}}\right)^{-\frac{2+\beta}{1+\beta}}\frac{\sigma_2/\sigma_1}{\Delta p^{2+\beta}}\right|$$

$$\Delta p \equiv |p_1 - p_{1c}| .$$

(3.18)

From here it is clear that near the percolation threshold, i.e., when $\Delta p<<1$ and $\sigma_2/\sigma_1<<1$, the leading part of σ_e/σ_1 is singular and has special scaling properties as function of those variables, namely,

(3.19) $$\sigma_e = \sigma_1 \Delta p^{1+\beta} F\left(\frac{\sigma_2/\sigma_1}{\Delta p^{2+\beta}}\right) .$$

The function F(z), which depends on only one variable, has simple analytic forms in three asymptotic regimes

$$F(z) \simeq \begin{cases} A + Bz + Cz^2 + \dots & z \ll 1,\ p_1 > p_{1c} \ :\ \text{regime I} \\ A'z + B'z^{\frac{2+\beta}{1+\beta}} + C'z^{\frac{3+\beta}{1+\beta}} + \dots & z \ll 1,\ p_1 < p_{1c} \ :\ \text{regime II} \\ A''z^{\frac{1+\beta}{2+\beta}} + B''z^{\frac{\beta}{2+\beta}} + C''z^{\frac{\beta-1}{2+\beta}} + \dots & z \gg 1,\ p_1 \lessgtr p_{1c} \ :\ \text{regime III} \end{cases} ,$$

(3.20)

where the coefficients A,A',A", etc. can easily be found from (3.17) and are all positive and $\mathcal{O}(1)$. The leading asymptotic forms of σ_e in the same regimes are, explicitly,

$$\sigma_e \simeq \begin{cases} \left(\frac{f(0)\ \Delta p}{p_{1c}^2\ a(1+\beta)}\right)^{1+\beta} \cdot \sigma_1 & \text{I} \\ \frac{b\ p_{1c}^2}{|f(\infty)|\Delta p} \cdot \sigma_2 & \text{II} \\ \left(\frac{b}{a(1+\beta)} \frac{f(0)}{|f(\infty)|}\right)^{\frac{1+\beta}{2+\beta}} \cdot \sigma_1^{\frac{1}{2+\beta}}\ \sigma_2^{\frac{1+\beta}{2+\beta}} & \text{III} \end{cases} .$$

(3.21)

In order to complete this calculation and make the results entirely explicit, we would have to determine the four constants f(0), $|f(\infty)|$, a,b. This requires a numerical evaluation of $\vec{E}(\vec{r})$ and the use of Eqs. (3.12) and (3.15), and will be done elsewhere.

4. Composite Conductor with a Weak Power Law Nonlinearity

We consider a composite conductor where the constitutive equation of each component has the form

$$(4.1)\quad \vec{J} = \sigma\vec{E} + b|E|^m \vec{E} \quad \text{for } b|E|^m \ll \sigma .$$

The linear conductivity σ and the nonlinearity coefficient b have different values in each component, but the exponent m is the same. Due to the smallness of the nonlinear term, it can be treated as a perturbation and everything can be expanded to first order in b. In this way it is easy to show that the volume averaged current density $\langle\vec{J}\rangle$ and electric field $\langle\vec{E}\rangle$ are related by [9]

$$(4.2)\quad \langle\vec{J}\rangle = \sigma_e \langle\vec{E}\rangle + b_e \left|\langle\vec{E}\rangle\right|^m \langle\vec{E}\rangle$$

where σ_e and b_e are given by

$$(4.3a)\quad \sigma_e = \frac{1}{V}\int dV\, \vec{\sigma}(\vec{r}) \left[\frac{\vec{E}(\vec{r})}{E_0}\right]^2$$

$$(4.3b)\quad b_e = \frac{1}{V}\int dV\, \vec{b}(\vec{r}) \left[\frac{\vec{E}(\vec{r})}{E_0}\right]^{m+2} .$$

$$\vec{E}_0 \equiv \langle\vec{E}\rangle$$

Note that the local field $\vec{E}(\vec{r})$ in these equations can be calculated by solving a purely linear problem, i.e., $\nabla\cdot\vec{J}=0$ where $\vec{J}(\vec{E})$ is given by (4.1) but without the nonlinear term.

In order to construct a Bruggeman-type approximation for the bulk effective nonelinearity coefficient b_e, we can again apply the procedure outlined in Section 2. In this case the calculation, though tedious, can be made entirely analytic since one only needs to find the distorted field of the linear problem when a uniform field is applied to a single spherical inclusion. If the host medium is characterized by the values σ_0,b_0 and the spherical inclusion of volume V_a is characterized by the values σ_i, b_i, we get the following expressions for σ_e, b_e in the case m=2 (cubic nonlinearity) and for general dimensionality d, correct up to and including $\mathcal{O}(V_a/V)$,

$$\epsilon_e = \sigma_0 + \sigma_0 \frac{V_a}{V} 3A_i$$

$$b_e = b_0 + \frac{V_a}{V} \{ b_0[-1 + 4A_i + \frac{2(d-1)(d+6)}{d+2}) A_i^2 +$$

$$+ \frac{4(d-1)(d-2)}{d+2} A_i^3 + \frac{1}{3} \frac{d-1}{d+2} (3d^2-7d + 6) A_i^4] + b_i(1-A_i)^4\}$$

$$A_i \equiv \frac{\sigma_i - \sigma_0}{\sigma_i + (d-1)\sigma_0} \quad .$$

(4.4)

The Bruggeman-type approximation is obtained, as usual, by equating the arithmetic average of the $\mathcal{O}(V_a/V)$ terms to zero. When this is done for b_e in a two-component composite, an approximation is obtained that is exact to $\mathcal{O}(1-p_1)$ for p_1 close to 1 and to $\mathcal{O}(p_1)$ for p_1 close to zero, and for any values of σ_1, σ_2. It is also exact to $\mathcal{O}(\sigma_1-\sigma_2)$ for $\sigma_1-\sigma_2 \ll \sigma_2$ and any value of p_1. But when we move away from those limits, the quality of the approximation deteriorates. This is most readily seen by considering the case of a metal-insulator composite where a percolation threshold is exhibited by σ_e at $p_{1c} = 1/d$. For $\sigma_2 = 0 < \sigma_1$, d=3, and arbitrary p_1 we then find [8]

$$\sigma_e = \begin{cases} \sigma_1 \cdot \frac{3}{2} \left(p_1 - \frac{1}{3}\right) & p_1 > \frac{1}{3} \\ 0 & p_1 < \frac{1}{3} \end{cases}$$

(4.5)

$$b_e = \begin{cases} b_1 \cdot \dfrac{405}{8} \dfrac{\left(p_1 - \frac{1}{3}\right)^4}{27p_1^3 - 18p_1^2 + 2p_1 - 1} & p_1 > 0.6412 \\ \infty & p_1 < 0.6412 \end{cases} \quad .$$

(4.6)

The result for b_e appears at first to be totally ridiculous, since an infinite value is obtained for all p_1 below some rather high value that is far above the percolation threshold ($p_{1c} = \frac{1}{3}$ in this approximation for σ_e). Further consideration shows that this is due to the fact that the integral in (4.3b) is very sensitive to local divergences of $\vec{E}(\vec{r})$. The results of (4.5) and (4.6), although obtained here by an uncontrolled approximation, are actually exact for a special type of microstructure that has an infinite hierarchy of length scales [10]. In this structure, $\vec{E}(\vec{r})$ is unbounded. This leads to no catastrophe when evaluating σ_e due to the variational properties of the integral in (4.3a). Those are responsible for the elementary bounds on σ_e, namely,

(4.7) $0 < \sigma_e < \sigma_1$.

By contrast, the integral in (4.3b) has no such variational properties and correspondingly no absolute upper bound exists for b_e. Of course, when $\vec{E}(\vec{r})$ diverges at some point, the use of (4.1) becomes unjustified because the inequality is violated. If we assume that (4.1) continues to hold even then, the perturbation expansion that lead to (4.2) breaks down. Nevertheless, the conclusion that there is no upper bound on b_e and that it becomes arbitrarily large remains correct.

We conclude from this that in order to find an approximation for b_e that is useful over the entire range of values of p_1, and in particular near a percolation threshold, some information about the microstructure must be incorporated. In particular, one would have to exclude the hierarchical type of microgeometry for which (4.5) and (4.6) are exact. Such approximations are currently being studied.

Acknowledgement:

This research was supported in part by the US-Israel Binational Science Foundation under Grant No. 88 - 00432.

References

[1] D.A.G. Bruggeman, Ann. Physik (Leipzig) **24** (1935), 636-664.

[2] D. Stroud, Phys. Rev. **B12** (1975), 3368-3373.

[3] D. Stroud and F. P. Pan, Phys. Rev. **B20** (1979), 455-465.

[4] S. Kirkpatrick, Rev. Mod. Phys. **45** (1973), 574-588.

[5] R. Hill, J. Mech. Phys. Solids **13** (1965), 213-222.

[6] B. Budiansky, J. Mech. Phys. Solids **13** (1965), 223-227.

[7] I. Webman, J. Jortner and M. H. Cohen, Phys. Rev. **B16** (1977), 2959-2964.

[8] D. J. Bergman, Phys. Rev. **B39** (1989), 4598-4609.

[9] D. Stroud and P. M. Hui, Phys. Rev. **B37** (1988), 8719-8724.

[10] G. Milton in Physics and Chemistry of Porous Media, eds. D. L. Johnson and P. N. Sen, AIP Conf. Proc. No. 107, New York, 1984, pp. 66-77.

David J. BERGMAN
School of Physics and Astronomy
Tel-Aviv University
Tel Aviv 69978, Israel

Homogenization of nonlinear unilateral problems

LUCIO BOCCARDO FRANÇOIS MURAT

1. Introduction

In this paper we consider the unilateral problems

$$(*)\qquad \begin{cases} u_\varepsilon \in K(\psi_\varepsilon) \\ <A_\varepsilon(u_\varepsilon) - f , v - u_\varepsilon> \geq 0 \\ \forall v \in K(\psi_\varepsilon) \end{cases}$$

where the right hand side f is fixed in $W^{-1,p'}(\Omega)$ and where the $A_\varepsilon(v)$ are monotone operators acting from $W_0^{1,p}(\Omega)$ into $W^{-1,p'}(\Omega)$ defined by

$$A_\varepsilon(v) = -\mathrm{div}(a_\varepsilon(x, Dv)) ,$$

while the unilateral convex sets $K(\psi_\varepsilon)$ are defined by

$$K(\psi_\varepsilon) = \{v \in W_0^{1,p}(\Omega) : v \geq \psi_\varepsilon \text{ a.e. in } \Omega\} .$$

We are concerned with the simultaneous variations of the nonlinear operators A_ε and of the obstacles ψ_ε : if the functions a_ε H-converge to a_0 and if the unilateral convex sets $K(\psi_\varepsilon)$ converge in the sense of Mosco to $K(\psi_0)$ (see Section 2 below for the definitions of these convergences) we prove that the solution u_ε of (*) converges to the solution of the same problem relative to $A_0(v) = -\mathrm{div}(a_0(x, Dv))$ and $K(\psi_0)$.

In the case where $p = 2$ and where the differential operators A_ε are linear, we proved the result in [BM]. In this paper, we also announced the "imminent" (see [K 1664]) publication of the analogous (present) result for monotone operators and also for higher order operators acting from $H_0^m(\Omega)$ into $H^{-m}(\Omega)$. Although that paper was never published, the results were presented on different occasions and in particular by the first of us at the conference "Equadiff 87" held in Xanthi (Greece).

At that time we confined ourselves to the case $p = 2$, since the only compactness result known for the homogenization of monotone operators was Tartar's result [T] which deals with the case $p = 2$. We were yet aware of the fact that our method worked in the general case, but the lack of compactness theorem for the homogenization of monotone operators when $p \neq 2$ discouraged us from publishing the result in this general case. The publication of the compactness result of V. Chiadò Piat, G. Dal Maso and A. Defranceschi [CDD] provided this setting and decided us to finally write the present paper.

To conclude this comment, and before turning to the presentation of the method we use, let us note that a result similar to the present one has been independently obtained by [DD] using different methods. Note also that the case where (*) is obtained through minimization of a convex functional on $K(\psi_\varepsilon)$ has been solved in [AP], [B2] using the Γ-convergence theory.

The method used here to pass to the limit in (*) is based on two arguments:

(i) the compactness lemma of [M2] which asserts that the injection of the positive cone of $W^{-1,p'}(\Omega)$ is compact in $W^{-1,q}(\Omega)$, for any $q < p'$: this result is here applied to $\mu_\varepsilon = A_\varepsilon u_\varepsilon - f$, which is positive since (*) is a unilateral problem.

(ii) the Meyer's regularity theorem ([Me], [MeE]) which asserts that the solutions $v_\varepsilon \in W_0^{1,p}(\Omega)$ of $A_\varepsilon v_\varepsilon = g$ are bounded in $W_0^{1,r}(\Omega)$, for some $r > p$, when g belongs to $W^{-1,s}(\Omega)$, $s > p'$, and the operators A_ε considered here are uniformly coercive and bounded. This regularity result, a density argument and Minty's trick allow us to compense the small loss of regularity (from $W^{-1,p'}(\Omega)$ to $W^{-1,q}(\Omega)$) which happens in (i).

This proof becomes simpler (see Subsection 4.1 below) in the case where a regularity theorem of Meyers' type ([B1]) is applied directly to the solution u_ε of the variational inequality (*).

Let us finally emphasize that our proof does not rely on any argument specific to second order equations. Indeed the same method based on (i) and (ii) immediately applies to monotone operators acting from $W_0^{m,p}(\Omega)$ into $W^{-m,p'}(\Omega)$ which H-converge, and to unilateral convex sets $K(\psi_\varepsilon)$ which converge in the sense of Mosco in $W_0^{m,p}(\Omega)$, allowing one to prove the corresponding convergence result. A new paper could be easily written in this case; we will only announce here the result, leaving its publication for the time where a theorem of compactness for monotone operators defined from $W_0^{m,p}(\Omega)$ into $W^{-m,p'}(\Omega)$ will be published.

Note however that the present proof is specific to the case of unilateral convex sets, because it strongly uses $\mu_\varepsilon \geq 0$ in (i). Indeed it is well known that the result would be false for different convex sets: consider e.g. the case of convex sets $K_\varepsilon = K$, where K is a compact subset of $W_0^{1,p}(\Omega)$.

2. Statements of the hypotheses and of the result

2.1 H-convergence

Let Ω be a bounded open subset of $\mathbb{R}^N$ (no smoothness is assumed on its boundary $\partial\Omega$) and let p, p' be real numbers such that

(2.1) $$1 < p, p' < +\infty \qquad \frac{1}{p} + \frac{1}{p'} = 1 .$$

Let ε be a parameter which takes its values in a sequence of strictly positive real numbers which tends to zero.

We consider here a sequence of nonlinear monotone operators defined by

(2.2) $$A_\varepsilon(v) = -\mathrm{div}(a_\varepsilon(x,Dv)) ,$$

where the functions a_ε are Carathéodory functions

$$(2.3) \qquad \begin{cases} a_\varepsilon : \Omega \times \mathbb{R}^N \to \mathbb{R}^N \\ a_\varepsilon \text{ measurable in } x \text{ and continuous in } \xi \end{cases}$$

which satisfy, for some constants $\alpha > 0$, $\beta > 0$ and $\gamma > 0$

$$(2.4) \qquad a_\varepsilon(x, \xi)\xi \geq \alpha|\xi|^p$$

$$(2.5) \qquad |a_\varepsilon(x, \xi)| \leq \beta|\xi|^{p-1} + \gamma$$

$$(2.6) \qquad [a_\varepsilon(x, \xi) - a_\varepsilon(x, \xi^*)]\,[\xi - \xi^*] > 0$$

for any ξ and ξ^* in $\mathbb{R}^N$, $\xi \neq \xi^*$ and for almost every x in Ω.

Under the assumptions (2.3), (2.4), (2.5), (2.6) the operator A_ε defined by (2.2) is a strictly monotone operator defined from $W_0^{1,p}(\Omega)$ into its dual $W^{-1,p'}(\Omega)$, and for any f in $W^{-1,p'}(\Omega)$ there exists a unique solution u_ε of the equation

$$(2.7) \qquad \begin{cases} -\mathrm{div}(a_\varepsilon(x, Du_\varepsilon)) = f \quad \text{in } \mathcal{D}'(\Omega) \\ u_\varepsilon \in W_0^{1,p}(\Omega)\,. \end{cases}$$

Moreover in view of the coerciveness condition (2.4), the solutions u_ε are bounded in $W_0^{1,p}(\Omega)$, independently of ε.

Definition 2.1. Consider a sequence of functions a_ε satisfying (2.3)-(2.6) and a function a_0 satisfying the same hypotheses (possibly with different constants $\alpha_0 > 0$, $\beta_0 > 0$ and $\gamma_0 > 0$). The sequence is said to *H-converge* to a_0 if for any f in $W^{-1,p'}(\Omega)$, the unique solution u_ε of (2.7) satisfies

$$(2.8) \qquad \begin{cases} u_\varepsilon \to u \quad \text{weakly in } W_0^{1,p}(\Omega) \\ a_\varepsilon(x, Du_\varepsilon) \to a_0(x, Du) \quad \text{weakly in } L^{p'}(\Omega)^N \end{cases}$$

where u is the unique solution of

$$(2.9)\qquad \begin{cases} -\mathrm{div}(a_0(x,Du)) = f \text{ in } \mathcal{D}'(\Omega) \\ u \in W_0^{1,p}(\Omega) . \end{cases}$$ ∎

This notion was first introduced by S. Spagnolo [S] under the name of G-convergence in the linear symmetric case, where $a_\varepsilon(x,\xi) = B_\varepsilon(x)\xi$ for some symmetric, coercive, $(L^\infty(\Omega))^{N^2}$ matrix B_ε (then $\gamma = 0$). S. Spagnolo proved in this case the following fundamental compactness theorem, which shows the relevance of the above definition: any sequence of symmetric, uniformly coercive and uniformly bounded matrices B_ε admits a subsequence which G-converges (or H-converges) to a matrix B_0 of the same class (defined by α, β).

These definition and compactness result were then generalized to the linear non symmetric case by L. Tartar and F. Murat [M1] and then by L. Tartar [T] to the case of nonlinear monotone operators of the form (2.2), defined from $H_0^1(\Omega)$ into $H^{-1}(\Omega)$, $(p = 2)$, when the functions a_ε satisfy

$$\begin{cases} [a_\varepsilon(x,\xi) - a_\varepsilon(x,\xi^*)]\,[\xi - \xi^*] \geq \alpha|\xi - \xi^*|^2 \\ |a_\varepsilon(x,\xi) - a_\varepsilon(x,\xi^*)| \leq \beta|\xi - \xi^*| \\ |a_\varepsilon(x,0)| \leq k(x) \in L^2(\Omega) . \end{cases}$$

Finally after some attempts by N. Fusco and G. Moscariello [FM] and by U. E. Raitum [R], a general result of compactness was recently proved by V. Chiadò Piat, G. Dal Maso and A. Defranceschi [CDD]. These authors consider the class of multivalued functions:

$$(2.10)\qquad \begin{cases} a_\varepsilon : \Omega \times \mathbb{R}^N \to \mathbb{R}^N \text{ measurable with closed values} \\ a_\varepsilon(x,\cdot) : \mathbb{R}^N \to \mathbb{R}^N \text{ maximal monotone multivalued function} \end{cases}$$

$$(2.11)\qquad \begin{cases} \sigma\xi \geq \bar{\alpha}|\xi|^p - \bar{a}(x) \\ \sigma\xi \geq \bar{\beta}|\sigma|^{p'} - \bar{b}(x) \\ \text{a.e. } x\in\Omega \ , \forall\xi\in \mathbb{R}^N \ , \forall\sigma\in a_\varepsilon(x,\xi) \end{cases}$$

where the constants $\bar{\alpha} > 0$, $\bar{\beta} > 0$ and the functions $\bar{a}\in L^1(\Omega)$, $\bar{b}\in L^1(\Omega)$ are given, and they write equation (2.7) in the form

$$(2.12)\qquad \begin{cases} -\mathrm{div}\sigma_\varepsilon = f \ \text{ in } \ \mathcal{D}'(\Omega) \\ \sigma_\varepsilon(x)\in a_\varepsilon(x,Du_\varepsilon(x)) \ \text{ a.e. } \ x\in\Omega \\ u_\varepsilon\in W_0^{1,p}(\Omega). \end{cases}$$

Extending the above definition of H-convergence ([CDD], Remark 3.10) they prove that any sequence of functions satisfying (2.10), (2.11) admits a subsequence which H-converges to a function a_0 of the same class.

2.2. Convergence in the sense of Mosco

In this paper we also consider unilateral closed convex sets $K(\psi)$ defined by

$$(2.13)\qquad K(\psi) = \{v\in W_0^{1,p}(\Omega) : v \geq \psi \ \text{ a.e. in } \ \Omega\} ,$$

where the obstacles ψ are measurable functions

$$\psi : \Omega \to \bar{\mathbb{R}} .$$

We only consider non empty sets $K(\psi)$.

Note that in (2.13) the definition of $K(\psi)$ is given using an almost everywhere inequality. When the obstacle ψ is a Borel function defined everywhere, a quasi everywhere inequality (for the $W_0^{1,p}(\Omega)$ capacity) can be equivalently used in place of the almost everywhere inequality.

Recall the following definition of convergence of convex sets due to U. Mosco [Mo].

Definition 2.2. Consider a sequence of non empty, closed, convex subsets C_0, C_ε of $W_0^{1,p}(\Omega)$. The sequence C_ε is said to converge to C_0 (in $W_0^{1,p}(\Omega)$) in the sense of Mosco if:

(i) for any v_0 in C_0 there exists a sequence $\bar{v}_\varepsilon$ such that

$$\text{(2.14)} \qquad \begin{cases} \bar{v}_\varepsilon \in C_\varepsilon \\ \bar{v}_\varepsilon \to v_0 \ \text{ strongly in } W_0^{1,p}(\Omega) \, ; \end{cases}$$

(ii) for any v and any subsequence η of ε such that

$$\text{(2.15)} \qquad \begin{cases} v_\eta \in C_\eta \\ v_\eta \to v \ \text{ weakly in } W_0^{1,p}(\Omega) \end{cases}$$

one has

$$\text{(2.16)} \qquad v \in C_0 \, . \qquad ∎$$

Numerous examples of non empty closed convex sets converging in the sense of Mosco can be found in the literature. For unilateral convex sets of type (2.13) the strong convergence

$$\psi_\varepsilon \to \psi_0 \ \text{ strongly in } W^{1,p}(\Omega)$$

easily implies the convergence of $K(\psi_\varepsilon)$ to $K(\psi_0)$, but the weak convergence

$$\psi_\varepsilon \to \psi_0 \ \text{ weakly in } W^{1,r}(\Omega) \, , \, r > p \, ,$$

also implies the same result (see [BM], [AP]). A necessary and sufficient condition for the convergence of $K(\psi_\varepsilon)$, expressed in terms of the convergence of the $W_0^{1,p}(\Omega)$ capacity of the level sets $\{x\in\Omega : \psi_\varepsilon(x) > t\}$ has been given in [D].

2.3. Main result

We can now state the main result of the present paper.

Theorem 2.3. *Consider a fixed right hand side*

$$f\in W^{-1,p'}(\Omega)\,, \tag{2.17}$$

a sequence of functions a_ε *satisfying (2.3)-(2.6) which* H*-converges to a function* a_0 *and a sequence* $K(\psi_\varepsilon)$ *of non empty unilateral convex sets of the type (2.13) which converges in* $W_0^{1,p}(\Omega)$ *in the sense of Mosco to* $K(\psi_0)$.

Under these hypotheses, the solution u_ε *of the unilateral problem*

$$\begin{cases} u_\varepsilon\in K(\psi_\varepsilon) \\ \displaystyle\int_\Omega a_\varepsilon(x,Du_\varepsilon)D(v-u_\varepsilon) \geq \langle f,v-u_\varepsilon\rangle \\ \forall v\in K(\psi_\varepsilon) \end{cases} \tag{2.18}$$

satisfies

$$\begin{cases} u_\varepsilon \to u_0 \quad \text{weakly in } W_0^{1,p}(\Omega) \\ a_\varepsilon(x,Du_\varepsilon) \to a_0(x,Du_0) \quad \text{weakly in } (L^{p'}(\Omega))^N \\ a_\varepsilon(x,Du_\varepsilon)Du_\varepsilon \to a_0(a,Du_0)Du_0 \quad \text{in } \mathcal{D}'(\Omega) \\ \displaystyle\int_\Omega a_\varepsilon(x,Du_\varepsilon)Du_\varepsilon \to \int_\Omega a_0(x,Du_0)Du_0 \end{cases} \tag{2.19}$$

where u_0 *is the solution of*

$$
(2.20) \qquad \begin{cases} u_0 \in K(\psi_0) \\ \displaystyle\int_\Omega a_0(x,Du_0)D(v - u_0) \geq \langle f, v - u_0 \rangle \\ \forall v \in K(\psi_0)\,. \end{cases}
$$

3. Proof of Theorem 2.3

As already noticed in the Introduction the present proof is based on the compactness in $W^{-1,q}_{loc}(\Omega)$ (for any $q < p'$) of the positive cone of $W^{-1,p'}(\Omega)$ (see [M2]) and on Meyers' regularity result ([Me], see also [G]). No argument relies here on the second order nature of the operator and the proof can be carried out as well in $W^{m,p}_0(\Omega)$ rather than in $W^{1,p}_0(\Omega)$, as done here.

Moreover the present proof does not really rely on the single valued and strictly monotone character of the functions a_ε and a_0 and could be adapted to obtain the result of Theorem 2.3 in the case where hypotheses (2.3)-(2.6) are replaced by hypotheses (2.10)-(2.11), just assuming the function $\overline{b}$ to belong to $L^\delta(\Omega)$ for some $\delta > 1$ in order to be able to use Meyers' regularity result.

In the case where the boundary $\partial\Omega$ is sufficiently smooth one can proceed on the whole of Ω and not only locally on Ω as done here. This simplifies the present proof because φ can be chosen equal to 1 on the whole of Ω in such case.

3.1. First step: a priori estimates

Existence and uniqueness of the solution of (2.18) is a classical result (see e.g. [KS]).

For a given v_0 in $K(\psi_0)$ consider the sequence $\overline{v}_\varepsilon$ which strongly converges in $W^{1,p}_0(\Omega)$ to v_0 in view of (2.14). Assumptions (2.4) and (2.5) easily imply that u_ε is bounded in $W^{1,p}_0(\Omega)$. Define

$$
(3.1) \qquad \begin{cases} \sigma_\varepsilon = a_\varepsilon(x,Du_\varepsilon) \\ \mu_\varepsilon = -\mathrm{div}(a_\varepsilon(x,Du_\varepsilon)) - f\,. \end{cases}
$$

Hypothesis (2.5) on a_ε yields the boundedness of σ_ε and μ_ε. We can thus extract a subsequence denoted by η such that, for some u^*, σ^*, μ^*,

(3.2) $$u_\eta \to u^* \text{ weakly in } W_0^{1,p}(\Omega) \text{ and strongly in } L^p(\Omega),$$

(3.3) $$\sigma_\eta \to \sigma^* \text{ weakly in } (L^{p'}(\Omega))^N,$$

(3.4) $$\mu_\eta \to \mu^* \text{ weakly in } W^{-1,p'}(\Omega).$$

Note that

(3.5) $$\mu^* = -\mathrm{div}\sigma^* - f$$

and

(3.6) $$u^* \in K(\psi_0)$$

because of (2.15) and (2.16).

3.2 Second step: compactness result in $W_{loc}^{-1,p'}(\Omega)$

Since $K(\psi_\varepsilon)$ is a unilateral convex set, we have (taking $v = u_\varepsilon + \varphi$ with $\varphi \in \mathcal{D}(\Omega)$, $\varphi \geq 0$ in (2.18)):

(3.7) $$\mu_\varepsilon \geq 0 \text{ in } \mathcal{D}'(\Omega).$$

In view of Theorem 1 of [M2], (3.4) and (3.7) imply that

(3.8) $$\mu_\eta \to \mu^* \quad \text{strongly in } W_{loc}^{1,q}(\Omega), \quad \forall q < p'.$$

3.3. Third step: proving that $\sigma^* = a_0(x,Du^*)$

Consider now a smooth function w_0, say

(3.9) $$w_0 \in \mathcal{D}(\Omega)$$

and the solution w_ε of the equation

(3.10)
$$\begin{cases} w_\varepsilon \in W_0^{1,p}(\Omega) \\ -\text{div}(a_\varepsilon(x,Dw_\varepsilon)) = -\text{div}(a_0(x,Dw_0)) \text{ in } \mathcal{D}'(\Omega) . \end{cases}$$

Meyers' regularity theorem (see [Me], [MeE] and [G]) implies that the solutions w_ε of (3.10) are uniformly bounded in $W^{1,r}_{loc}(\Omega)$ for some $r > p$, since the functions a_ε satisfy (2.3)-(2.6) (it will actually be sufficient for γ in (2.5) to belong to $L^s(\Omega)$ with $s > p'$ in order to have Meyers' regularity result). To be more precise, for any bounded open subset ω such that

$$\bar{\omega} \subset \Omega ,$$

there exists some $r > p$ and some constant c (which only depend on Ω, ω and on α, β and γ which appear in the hypotheses (2.3)-(2.6)) such that

(3.11)
$$\|w_\varepsilon\|_{W^{1,r}(\omega)} \le c .$$

Note that r and c can be chosen independently of ω when $\partial\Omega$ is sufficiently smooth. In such case the choice $\omega = \Omega$ becomes licit.

Fix now φ and an open subset ω such that

(3.12)
$$\varphi \in \mathcal{D}(\Omega) , 0 \le \varphi \le 1 , \varphi = 0 \text{ on } \Omega \setminus \bar{\omega} .$$

Note that in the case where $\partial\Omega$ is sufficiently smooth, φ can be fixed equal to 1 on Ω. This simplifies the rest of the present proof and allows one to recover any result proved in this Section except the results (3.25) and (3.30) which are concerned with the local convergence of the energy.

Combined with Definition 2.1, the bound (3.11) implies that

(3.13)
$$\begin{cases} w_\varepsilon \to w_0 \text{ weakly in } W^{1,r}(\omega) \\ a_\varepsilon(x,Dw_\varepsilon) \to a_0(x,Dw_0) \text{ weakly in } (L^{p'}(\omega))^N . \end{cases}$$

Consider now the inequality

(3.14) $$\int_\Omega \varphi[a_\varepsilon(x,Du_\varepsilon) - a_\varepsilon(x,Dw_\varepsilon)]\,[Du_\varepsilon - Dw_\varepsilon] \geq 0\,.$$

Since u^* belongs to $K(\psi_0)$, there exists a sequence $\bar{u}_\varepsilon$ (see (2.14)) such that

(3.15) $$\begin{cases} \bar{u}_\varepsilon \to u^* \text{ strongly in } W_0^{1,p}(\Omega) \\ u_\varepsilon \in K(\psi_\varepsilon)\,. \end{cases}$$

Using $\bar{u}_\varepsilon$ we rewrite (3.13) as

(3.16) $$\begin{cases} I_\varepsilon + II_\varepsilon + III_\varepsilon + IV_\varepsilon \geq 0 \\ I_\varepsilon = \int_\Omega \varphi\sigma_\varepsilon[Du_\varepsilon - D\bar{u}_\varepsilon] \\ II_\varepsilon = \int_\Omega \varphi\sigma_\varepsilon\, D\bar{u}_\varepsilon \\ III_\varepsilon = -\int_\Omega \varphi\sigma_\varepsilon\, Dw_\varepsilon \\ IV_\varepsilon = -\int_\Omega \varphi a_\varepsilon(x,Dw_\varepsilon)[Du_\varepsilon - Dw_\varepsilon]\,. \end{cases}$$

Since $0 \leq \varphi \leq 1$ on Ω , the function v defined by

$$v = \varphi\bar{u}_\varepsilon + (1-\varphi)\, u_\varepsilon$$

belongs to $K(\psi_\varepsilon)$ and can be used as test function in (2.18) giving

(3.17) $$\int_\Omega \sigma_\varepsilon\, D[\varphi(\bar{u}_\varepsilon - u_\varepsilon)] \;\geq\; \langle f, \varphi(\bar{u}_\varepsilon - u_\varepsilon)\rangle\,.$$

This implies that

$$I_\varepsilon = \int_\Omega \sigma_\varepsilon D[\varphi(u_\varepsilon - \bar{u}_\varepsilon)] - \int_\Omega (u_\varepsilon - \bar{u}_\varepsilon)\sigma_\varepsilon D\varphi$$

$$\leq \langle f, \varphi(u_\varepsilon - \bar{u}_\varepsilon)\rangle - \int_\Omega (u_\varepsilon - \bar{u}_\varepsilon)\sigma_\varepsilon D\varphi .$$

Since u_η and $\bar{u}_\varepsilon$ tend to u^* weakly in $W_0^{1,p}(\Omega)$ and strongly in $L^p(\Omega)$, we have

(3.18) $$\limsup_{\eta \to 0} I_\eta \leq 0 .$$

The strong convergence of $\bar{u}_\varepsilon$ to u^* in $W_0^{1,p}(\Omega)$ yields:

(3.19) $$II_\varepsilon \to \int_\Omega \varphi\sigma^* Du^* .$$

Rewrite III_ε as

$$III_\varepsilon = \langle \operatorname{div} \sigma_\varepsilon , \varphi w_\varepsilon \rangle + \int_\Omega w_\varepsilon \sigma_\varepsilon D\varphi .$$

Since $\mu_\eta = -\operatorname{div} \sigma_\eta - f$ strongly converges in $W_{loc}^{-1,q}(\Omega)$ for any $q < p'$ (see (3.8) and (3.5)) while φw_ε tends to φw_0 weakly in $W_0^{1,r}(\omega)$, $r > p$ (see (3.13)), it is easy to pass to the limit in the first term of III_ε. For the second one it is sufficient to use (3.3) and (3.13). This yields

(3.20) $$III_\eta \to \langle \operatorname{div} \sigma^* , \varphi w_0 \rangle + \int_\Omega w_0 \sigma^* D\varphi = - \int_\Omega \varphi\sigma^* Dw_0 .$$

Similarly we have

$$IV_\varepsilon = \langle \operatorname{div}(a_0(x,Dw_0)) , \varphi(u_\varepsilon - w_\varepsilon)\rangle + \int_\Omega (u_\varepsilon - w_\varepsilon) a_\varepsilon(x,Dw_\varepsilon) D\varphi ,$$

which passing to the limit yields

(3.21)
$$IV_\eta \longmapsto -\int_\Omega \varphi a_0(x,Dw_0)\,[Du^* - Dw_0]\,.$$

Combining (3.16), (3.18), (3.19), (3.20) and (3.21) we have

(3.22)
$$\int_\Omega \varphi[\sigma^* - a_0(x,Dw_0)]\,[Du^* - Dw_0] \geq 0\,.$$

This inequality was proved for any w_0 in $\mathcal{D}(\Omega)$. By density (3.22) remains valid for any w_0 in $W_0^{1,p}(\Omega)$. Consider $w_0 = u^* + t\varphi$, with $t \in \mathbb{R}^+$ and $\varphi \in W_0^{1,p}(\Omega)$. Letting t tend to zero we obtain by Minty's trick

(3.23)
$$\sigma^* = a_0(x,Du^*)\,.$$

3.4. Fourth step: lower semicontinuity of the energy

Returning to (3.14), (3.16) we have

$$\int_\Omega \varphi\sigma_\varepsilon\, Du_\varepsilon \geq \int_\Omega \varphi\sigma_\varepsilon\, Dw_\varepsilon + \int_\Omega \varphi a_\varepsilon(x,Dw_\varepsilon)\,[Du_\varepsilon - Dw_\varepsilon]$$

$$= -III_\varepsilon - IV_\varepsilon\,.$$

Using (3.20), (3.21), (3.23) we have for any w_0 in $\mathcal{D}(\Omega)$

(3.24)
$$\begin{cases} \liminf\limits_{\eta\to 0} \displaystyle\int_\Omega \varphi\sigma_\eta Du_\eta \\ \geq \displaystyle\int_\Omega \varphi a_0(x,Du^*)\, Dw_0 + \int_\Omega \varphi a_0(x,Dw_0)[Du^* - Dw_0]\,. \end{cases}$$

Letting w_0 tend to u^* in $W_0^{1,p}(\Omega)$ we obtain

$$(3.25)\qquad \begin{cases} \liminf\limits_{\eta\to 0} \displaystyle\int_\Omega \varphi\sigma_\eta Du_\eta \geq \int_\Omega \varphi a_0(x,Du^*)\, Du^* \\ \forall \varphi \in \mathcal{D}(\Omega)\ ,\ 0 \leq \varphi \leq 1\ . \end{cases}$$

3.5. Fifth step: u^* is the solution u_0 of (2.20)

Consider now some v_0 in $K(\psi_0)$ and the sequence $\bar{v}_\varepsilon$ associated to v_0 by (2.14). Using $\bar{v}_\varepsilon$ as test function in (2.18) as well as a function $\varphi \in \mathcal{D}(\Omega)$, $0 \leq \varphi \leq 1$, we obtain

$$(3.26)\qquad \int_\Omega \sigma_\varepsilon D\bar{v}_\varepsilon - \langle f, \bar{v}_\varepsilon - u_\varepsilon\rangle \geq \int_\Omega \sigma_\varepsilon Du_\varepsilon \geq \int_\Omega \varphi\sigma_\varepsilon Du_\varepsilon\ .$$

It is easy to pass to limit in η in each term of (3.26) using (3.25), (2.14), (3.3), (3.23). This implies

$$(3.27)\qquad \begin{cases} \displaystyle\int_\Omega a_0(x,Du^*)Dv_0 - \langle f, v_0 - u^*\rangle \geq \int_\Omega \varphi a_0(x,Du^*)Du^* \\ \forall v_0 \in K(\psi_0)\ ,\ \forall \varphi \in \mathcal{D}(\Omega)\ ,\ 0 \leq \varphi \leq 1\ . \end{cases}$$

Passing to the limit in the right hand side of (3.27) with φ tending to 1 then gives

$$(3.28)\qquad \begin{cases} \displaystyle\int_\Omega a_0(x,Du^*)\,(Dv_0 - Du^*) - \langle f, v_0 - u^*\rangle \geq 0 \\ \forall v_0 \in K(\psi_0)\ . \end{cases}$$

Since u^* belongs to $K(\psi_0)$ (see (3.6)), u^* actually coincides with the unique solution u_0 of (2.20). This implies that the whole sequences u_ε and $a_\varepsilon(x,Du_\varepsilon)$ (and not only subsequences) converge to u_0 and $a_0(x,Du_0)$.

3.6. Sixth step: convergence of the energy

We first deduce from (3.17) that for any $\varphi \in \mathcal{D}(\Omega)$, $0 \leq \varphi \leq 1$

$$\int_\Omega \varphi\sigma_\varepsilon Du_\varepsilon \leq \int_\Omega \sigma_\varepsilon D(\varphi\bar{u}_\varepsilon) - \int_\Omega u_\varepsilon \sigma_\varepsilon D\varphi - \langle f, \varphi(\bar{u}_\varepsilon - u_\varepsilon)\rangle$$

which implies

$$\limsup_{\varepsilon\to 0} \int_\Omega \varphi\sigma_\varepsilon Du_\varepsilon \leq \int_\Omega \varphi a_0(x,Du_0)\, Du_0 \,. \tag{3.29}$$

Combined with (3.25), inequality (3.29) implies that

$$\sigma_\varepsilon\, Du_\varepsilon \to a_0(x,Du_0)Du_0 \quad \text{in } \mathcal{D}'(\Omega)\,. \tag{3.30}$$

On the other hand, taking $v_0 = u_0$ in (3.26) we obtain from (2.14) and (3.26) that

$$\int_\Omega a_0(x,Du_0)Du_0 \geq \limsup_{\varepsilon\to 0} \int_\Omega \sigma_\varepsilon Du_\varepsilon$$

$$\geq \liminf_{\varepsilon\to 0} \int_\Omega \sigma_\varepsilon Du_\varepsilon \geq \int_\Omega \varphi a_0(x,Du_0)Du_0 \,,$$

for any $\varphi \in \mathcal{D}(\Omega)$, $0 \leq \varphi \leq 1$, which implies, letting φ tend to 1, that

$$\int_\Omega a_\varepsilon(x,Du_\varepsilon)Du_\varepsilon \to \int_\Omega a_0(x,Du_0)Du_0 \,. \tag{3.31}$$

This completes the proof of Theorem 2.3.

4. Strong convergence of $A_\varepsilon u_\varepsilon$ in $W^{-1,p'}(\Omega)$

In this Section we prove that under some strong monotonicity assumptions one has

(4.1) $$-\operatorname{div}(a_\varepsilon(x,Du_\varepsilon)) \to -\operatorname{div}(a_0(x,Du_0)) \text{ strongly in } W^{-1,p'}(\Omega)$$

(compare with the weak convergence obtained in (2.19)).
This strong convergence can be used to obtain for the solution of the unilateral problem (2.18) a corrector result similar to the one obtained for the equation (see [M1], [T]).

In the proof we use an extension of Meyers' regularity result [B1] which deals with the solution of the unilateral problem itself. We begin (Subsection 4.1) with the case where the obstacle is fixed and where some regularity assumptions are met. In such case the proof is very simple.

In the whole of this Section we will assume that

(4.2) $$\partial\Omega \text{ is sufficiently smooth;}$$

and in Subsection 4.2 that

(4.3) $$\psi_0 \in W_0^{1,p}(\Omega)$$

(4.4a) $$\begin{cases} [a_\varepsilon(x,\xi) - a_\varepsilon(x,\xi^*)]\,[\xi-\xi^*] \ge \alpha|\xi-\xi^*|^p , & \text{if } p \ge 2 \\ [a_\varepsilon(x,\xi) - a_\varepsilon(x,\xi^*)]\,[\xi-\xi^*] \ge \alpha \dfrac{|\xi-\xi^*|^2}{(|\xi| + |\xi^*|)^{2-p}} , & \text{if } p \le 2 \end{cases}$$

and in Subsection 4.2 that

(4.4b) $$\begin{cases} |a_\varepsilon(x,\xi) - a_\varepsilon(x,\xi^*)| \le \beta(|\xi| + |\xi^*|)^{p-2}|\xi-\xi^*| & \text{if } p \ge 2 \\ |a_\varepsilon(x,\xi) - a_\varepsilon(x,\xi^*)| \le \beta|\xi-\xi^*|^{p-1} & \text{if } p \le 2 . \end{cases}$$

4.1. First step: fixed smooth obstacle

Consider some obstacle $\hat{\psi}$ and right hand side $\hat{f}$ which satisfy

$$(4.5)\qquad \begin{cases} \hat{\psi} \in W_0^{1,s}(\Omega) \ , \ s > p \\ \hat{f} \in W^{-1,r}(\Omega) \ , \ r > p' . \end{cases}$$

Consider also a sequence of functions a_ε which satisfy (2.3)-(2.6) and which H-converge to a function a_0 satisfying (2.3)-(2.6) (possibly with different constants α_0, β_0 and γ_0).

Define the unilateral problems

$$(4.6)\qquad \begin{cases} \hat{u}_\varepsilon \in K(\hat{\psi}) \\ \int_\Omega a_\varepsilon(x, D\hat{u}_\varepsilon)(Dv - D\hat{u}_\varepsilon) \ \geq \ <\hat{f}, v - \hat{u}_\varepsilon> \\ \forall v \in K(\hat{\psi}) . \end{cases}$$

Defining $\hat{\mu}_\varepsilon$ as

$$(4.7)\qquad \hat{\mu}_\varepsilon = -\text{div}(a_\varepsilon(x, D\hat{u}_\varepsilon)) - \hat{f} ,$$

the unilateral problem (4.6) is known to be equivalent to the "complementarity system"

$$(4.8)\qquad \begin{cases} \hat{u}_\varepsilon \geq \hat{\psi} \\ \hat{\mu}_\varepsilon \geq 0 \\ <\hat{\mu}_\varepsilon, \hat{u}_\varepsilon - \hat{\psi}> = 0 . \end{cases}$$

Let us now invoke the extension of Meyers' regularity theorem ([B1]) which asserts that under the present hypotheses the solutions $\hat{u}_\varepsilon$ of (4.6) are bounded in $W_0^{1,\sigma}(\Omega)$, for some $\sigma > p$ (this number σ depends on Ω, $\hat{f}$, $\hat{\psi}$, α, β and γ); then $\hat{\mu}_\varepsilon$ are bounded in $W^{-1,\rho}(\Omega)$, for some $\rho > p'$.

Since we have $\hat{\mu}_\varepsilon \geq 0$, the compactness embedding of the positive cone of $W^{-1,p}(\Omega)$ into $W^{-1,p'}(\Omega)$ ([M2]) (which is valid on the whole of Ω) implies that $\hat{\mu}_\varepsilon$ is relatively compact in $W^{-1,p'}(\Omega)$. We can thus extract a subsequence $\hat{u}_\eta$ such that, for some $\hat{u}$ and $\hat{\mu}$,

(4.9) $$\hat{u}_\eta \to \hat{u} \text{ weakly in } W_0^{1,p}(\Omega)$$

(4.10) $$\hat{\mu}_\eta \to \hat{\mu} \text{ strongly in } W^{-1,p'}(\Omega).$$

Considering (4.7) as an equation on $\hat{u}_\varepsilon$ with right hand side $\hat{f} + \hat{\mu}_\varepsilon$, one can prove from the definition 2.1 of H-convergence that

(4.11) $$\hat{\mu} = -\mathrm{div}(a_0(x, D\hat{u})) - \hat{f} :$$

the proof is straightforward if (4.4a) is assumed to hold (see (4.21) below when $p < 2$) but can be as well carried aut without assuming this hypothesis.
Passing to the limit in (4.8) we have

(4.12) $$\begin{cases} \hat{u} \geq \hat{\psi} \\ \hat{\mu} \geq 0 \\ <\hat{\mu}, \hat{u} - \hat{\psi}> = 0 \end{cases}$$

which implies that $\hat{u}$ is the (unique) solution of the unilateral problem

(4.13) $$\begin{cases} \hat{u} \in K(\hat{\psi}) \\ \int_\Omega a_0(x, D\hat{u})(Dv - D\hat{u}) \geq <\hat{f}, v - \hat{u}> \\ \forall v \in K(\hat{\psi}) ; \end{cases}$$

therefore the whole sequences (and not only subsequences) converge in (4.9), (4.10).

Note that (4.10) immediately implies (4.1) in the present setting.

4.2. Second step: general case

Let us now consider both the unilateral problems (2.18) and (4.6) and write

$$\int_\Omega [a_\varepsilon(x,Du_\varepsilon) - a_\varepsilon(x,D\hat{u}_\varepsilon)]\,[Du_\varepsilon - D\hat{u}_\varepsilon] = I_\varepsilon + II_\varepsilon + III_\varepsilon\,, \tag{4.14}$$

where

$$\begin{cases} I_\varepsilon = \langle -\mathrm{div}(a_\varepsilon(x,Du_\varepsilon)) - f, u_\varepsilon - \hat{u}_\varepsilon\rangle \\ II_\varepsilon = \langle f - \hat{f}, u_\varepsilon - \hat{u}_\varepsilon\rangle \\ III_\varepsilon = \langle \hat{f} + \mathrm{div}(a_\varepsilon(x,D\hat{u}_\varepsilon)), u_\varepsilon - \hat{u}_\varepsilon\rangle\,. \end{cases} \tag{4.15}$$

Extracting a subsequence u_η, we have (see (3.2), (3.4))

$$\begin{cases} u_\eta \to u^* \text{ weakly in } W_0^{1,p}(\Omega) \\ -\mathrm{div}(a_\eta(x,Du_\eta)) - f \to \mu^* \text{ weakly in } W^{-1,p'}(\Omega)\,, \end{cases}$$

and in view of (4.9), (4.10) and (4.11), it is easy to pass to the limit in II_ε and III_ε:

$$\begin{cases} II_\eta \to \langle f - \hat{f}, u^* - \hat{u}\rangle \\ III_\eta \to \langle -\hat{\mu}, u^* - \hat{u}\rangle = \langle \hat{f} + \mathrm{div}(a_0(x,D\hat{u})), u^* - \hat{u}\rangle\,. \end{cases} \tag{4.16}$$

For what concerns I_ε, let us introduce $\bar{\psi}_\varepsilon$ which satisfies

$$\begin{cases} \bar{\psi}_\varepsilon \in K(\psi_\varepsilon) \\ \bar{\psi}_\varepsilon \to \psi_0 \text{ strongly in } W_0^{1,p}(\Omega)\,; \end{cases}$$

the existence of such a sequence $\bar{\psi}_\varepsilon$ is asserted by (2.14) since ψ_0 belongs to $W_0^{1,p}(\Omega)$ (see hypothesis (4.3)). Since $\hat{u}_\varepsilon - \hat{\psi}$ is positive and belongs to $W_0^{1,p}(\Omega)$ we can use $v = \bar{\psi}_\varepsilon + (\hat{u}_\varepsilon - \hat{\psi})$ as test function in (2.18), obtaining

$$<-\mathrm{div}(a_\varepsilon(x,Du_\varepsilon)) - f\,,\ \bar{\psi}_\varepsilon + (\hat{u}_\varepsilon - \hat{\psi}) - u_\varepsilon> \ \geq\ 0\,.$$

This implies that

$$I_\varepsilon \leq <-\mathrm{div}(a_\varepsilon(x,Du_\varepsilon)) - f,\ \bar{\psi}_\varepsilon - \hat{\psi}>$$

which yields

$$\limsup_{\eta\to 0} I_\eta \leq <\mu^*, \psi_0 - \hat{\psi}>\,. \tag{4.17}$$

Defining

$$R(\hat{\psi}, \hat{f}) = <\mu^*, \psi_0 - \hat{\psi}> + <f - \hat{f}, u^* - \hat{u}> + <\hat{f} + \mathrm{div}(a_0(x, D\hat{u})), u^* - \hat{u}> \tag{4.18}$$

we have proved that

$$\limsup_{\eta\to 0} \int_\Omega [a_\eta(x,Du_\eta) - a_\eta(x,D\hat{u}_\eta)][Du_\eta - D\hat{u}_\eta] \leq R(\hat{\psi}, \hat{f})\,. \tag{4.19}$$

Since we assumed the strong monotonicity hypothesis (4.4a), we claim that (4.19) implies that

$$\alpha \|u_\eta - \hat{u}_\eta\|^p_{W_0^{1,p}(\Omega)} \leq c\, R(\hat{\psi}, \hat{f})\,, \tag{4.20}$$

where c is a constant depending on the data such that $c = 1$ if $p \geq 2$. When $p \geq 2$ assertion (4.20) is immediate. When $1 < p \leq 2$, Hölder's inequlity yields for any w, z in $W_0^{1,p}(\Omega)$

$$(4.21)\quad \begin{cases} \displaystyle\int_\Omega |D(w-z)|^p = \int_\Omega \frac{|D(w-z)|^p}{|Dw|+|Dz|^{p(2-p)/2}} (|Dw|+|Dz|)^{p(2-p)/2} \\ \displaystyle\leq [\int_\Omega \frac{|D(w-z)|^2}{(|Dw|+|Dz|)^{2-p}}]^{p/2} [\int_\Omega (|Dw|+|Dz|)^p]^{(2-p)/2} . \end{cases}$$

Inequality (4.20) thus follows from (4.21), (4.4a) and from the a priori estimates on the solutions u_η and $\hat{u}_\eta$.

Let now $\hat{\psi}$ and $\hat{f}$ be smooth approximations of ψ_0 and f, i.e. let us assume that

$$(4.22)\quad \begin{cases} \hat{\psi} \to \psi_0 \text{ strongly in } W_0^{1,p}(\Omega) \\ \hat{f} \to f \text{ strongly in } W^{-1,p'}(\Omega) . \end{cases}$$

In such case it can be proved (this is easy in the present setting using hypothesis (4.4a)) that the solution $\hat{u}$ of (4.13) strongly converges in $W_0^{1,p}(\Omega)$ to the solution u_0 of (2.20) and thus

$$(4.23)\quad R(\hat{\psi}, \hat{f}) \to \langle f + \mathrm{div}(a_0(x,Du_0)), u^* - u_0\rangle \leq 0$$

which is non-positive since u^* belongs to $K(\psi_0)$ because of (2.15), (2.16).

From (4.20), (4.23) and from the result of the first step, it is easy to prove that $u^* = u_0$, i.e. that u_η weakly converges in $W_0^{1,p}(\Omega)$ to the solution u_0 of (2.20); this implies that the hole sequences (and not only subsequences) converge. Moreover (4.20), (4.23), hypothesis (4.4) and (4.4b), the strong convergence (4.10) of $\hat{\mu}_\eta$ to $\hat{\mu}$ and the strong convergence of $\hat{u}$ to u_0 imply that

$$(4.24)\quad -\mathrm{div}(a_\varepsilon(x,Du_\varepsilon)) \to -\mathrm{div}(a_0(x,Du_0)) \text{ strongly in } W^{-1,p'}(\Omega)$$

which is the desired result (4.1).

Acknowledgements

The present paper contains part of the results presented by the first author at the conference "Composite media and homogenization theory" (Trieste, January 1990). Both authors would like to thank the organizers of this conference for having given them the opportunity of presenting their work and of writing the present paper. Other results (concerned with homogenization of quasilinear equations involving lower order terms which are nonlinear with respect to the gradient) will be published elsewhere ([BBM]).

REFERENCES

[AP] H. Attouch, C. Picard, Variational inequalities with varying obstacles. The general form of the limit problem, J. Funct. Anal. **50** (1983), 329-386.

[B1] L. Boccardo, An L^s-estimate for the gradient of solutions of some nonlinear unilateral problems, Ann. Mat. Pura Appl. **141** (1985), 277-287.

[B2] L. Boccardo, L^∞ and L^1 variations on a theme of Γ-convergence, in Partial differential equations and the calculus of variations, Essays in honour of Ennio De Giorgi Vol. 1, F. Colombini et al. ed., Birkhäuser (1989), 135-147.

[BBM] A. Bensoussan, L. Boccardo, F. Murat, H-convergence for quasilinear elliptic equations with quadratic growth, to appear.

[BM] L. Boccardo, F. Murat, Nouveaux résultats de convergence dans des problèmes unilateraux, in Nonlinear partial differential equations and their applications, Collège de France seminar, Vol. II, H. Brezis, J.-L. Lions ed., Research Notes in Math. **60**, Pitman (1982), 64-85.

[CDD] V. Chiadò Piat, G. Dal Maso, A. Defranceschi, G-convergence of monotone operators, Ann. Inst. H. Poincaré, Analyse non Linéaire, to appear.

[D] G. Dal Maso, Some necessary and sufficient conditions for the convergence of sequences of unilateral convex sets, J. Funct. Anal. **62** (1985), 119-159.

[DD] G. Dal Maso, A. Defranceschi, Convergence of unilateral problems for monotone operators, J. Analyse Math. **53** (1989), 269-289.

[FM] N. Fusco, G. Moscariello, On the homogenization of quasilinear divergence structure operators, Ann. Mat. Pura Appl. **146** (1987), 1-13.

[G] M. Giaquinta, Multiple integrals in the calculus of variations and nonlinear elliptic systems, Annals of Math. Studies **105**, Princeton (1983).

[K 1664] Kronenburg (1664): La recherche de la perfection dans la fabrication de cette bière la rend digne de trois siècles de tradition.

[KS] D. Kinderlehrer, G. Stampacchia, An introduction to variational inequalities and their applications, Academic Press (1980).

[M1] F. Murat, H-convergence - Seminar d'analyse fonctionelle et numérique, Université d'Alger, 1977-78, multigraphed.

[M2] F. Murat, L'injection du cône positif de H^{-1} dans $W^{-1,q}$ est compacte pour tout $q < 2$, J. Math. Pures et Appl. **60** (1981), 309-322.

[Me] N.G. Meyers, An L^p-estimate for the gradient of solutions of second order elliptic divergence equations, Ann. Sc. Norm. Sup. Pisa **17** (1963), 189-206.

[MeE] N.G. Meyers, A. Elcrat, Some results on regularity for solutions of nonlinear elliptic systems and quasiregular functions, Duke Math. J. **42** (1975), 121-136.

[Mo] U. Mosco, Convergence of convex sets and of solutions of variational inequalities, Adv. in Math. **3** (1969), 510-585.

[R] U.E. Raitum, On the G-convergence of quasi-linear elliptic operators with unbounded coefficients, Sov. Math. Dokl. **24** (1981), 472-475.

[S] S. Spagnolo, Sulla convergenza di soluzioni di equazioni paraboliche ed ellittiche, Ann. Sc. Norm. Sup. Pisa **22** (1968), 517-597.

[T] L. Tartar, Cours Peccot au Collège de France, March 1977.

Lucio Boccardo
Dipartimento di Matematica
Università di Roma I
Piazzale Aldo Moro, 2
00185 ROMA,
ITALY

François Murat
Laboratoire d'Analyse Numérique
Université Paris VI
4, Place Jussieu
75252 PARIS CEDEX 05
FRANCE

HOMOGENIZATION, PLASTICITY AND YIELD DESIGN

G. BOUCHITTE   P. SUQUET

Abstract: We consider an epi-convergence problem arising from the theory of yield design. The functional under consideration has a linear growth with respect to the deformation tensor of the displacement field, and the problem is naturally posed in a space of displacement fields with bounded deformation. The problem includes a linear constraint which can be closed or not closed, depending on the type of boundary conditions considered. In the case where the constraint is not closed (applied forces on a part of the boundary) a relaxation term appears. Physically the strength of the loaded boundary turns out to be smaller than the natural guess deduced from the well known Average Variational Principle.

1. OUTLINE OF THE PAPER

This paper deals with an homogenization problem arising from *yield design*, i.e. from the mechanical theory which predicts the load carrying capacity of a structure made from materials with a limited strength. In its dual form, the yield design problem (often called the limit load problem) for a finely periodic structure can be written as a variational problem:

$$(1.1) \qquad \lambda^{\epsilon} = \inf_{\substack{u=0 \text{ sur } \Gamma_0 \\ L(u)=1}} J^{\epsilon}(u) = \int_{\Omega} j^{\epsilon}(x,e(u))\,dx \quad .$$

where Ω is a bounded open set in $\mathbb{R}^N$, $u:\Omega \to \mathbb{R}^N$ is a vector valued field (rate of displacement), $e(u)$ is its deformation tensor, Γ_0 is a subset of the boundary $\partial\Omega$, $L(u)$ is the linear form:

$$(1.2) \qquad L(u) = \int_{\Omega} f_0 . u\,dx + \int_{\Gamma_1} g_0 . u\,ds \quad .$$

where $\Gamma_1 = \partial\Omega - \Gamma_0$.

$j^{\epsilon}(x,E)$ is ϵ-periodic with respect to the variable x, and is positively homogeneous of degree one with respect to the variable E:

(1.3) $$j^{\epsilon}(x+\epsilon T,E) = j^{\epsilon}(x,E) \text{ for every } T \text{ in } \mathbb{Z}^N,$$

(1.4) $$j^{\epsilon}(x,\lambda E) = \lambda\, j^{\epsilon}(x,E) \text{ for every } E \text{ in } \mathbb{R}_s^{N^2} \text{ and every } \lambda > 0.$$

Our objective is to find the limit of λ^{ϵ} and of J^{ϵ} (in the sense of Γ-limits) when ϵ goes to 0.

A first guess for this limit, inspired by the *Average Variational Principle* (A.V.P.) (BENSOUSSAN&al [4], MARCELLINI [14]), would be to replace j^{ϵ} in (1.1) by j^{hom} defined by a variational problem on the unit cell $Y =]0,1[^N$, which generates the entire geometry by periodicity:

(1.5) $$\lambda^0 = \operatorname*{Inf}_{\substack{u = 0 \text{ on } \Gamma_0 \\ L(u)=1}} \int_{\Omega} j^{hom}(e(u))\, dx ,$$

(1.6) $$j^{hom}(E) = \operatorname*{Inf}_{w \text{ periodic}} \frac{1}{|Y|} \int_Y j(y,E+e(w))\, dy .$$

This first guess is correct *if $\Gamma_0 = \partial\Omega$, or if g vanishes identically* (i.e. if the boundary is not loaded by imposed external forces). More specifically in this case:

(1.7) $$\lim_{\epsilon \to 0} \lambda^{\epsilon} = \lambda^0 = \operatorname*{Inf}_{L(u)=1} \int_{\Omega} j^{hom}(e(u))\, dx + \int_{\Gamma_0} j^{hom}(-u \otimes_s n)\, ds.$$

The second integral in (1.7) is a classical relaxation term associated to the first integral term, and accounts for the loss of the boundary condition $u|_{\Gamma_0} = 0$. This term is classical in the theory of minimal surfaces, and in Plasticity TEMAM [17].

Surprisingly, when the boundary is loaded ($\text{mes}(\Gamma_1)>0$, $g \not\equiv 0$) the preceeding guess (1.6) overestimates the exact result, SUQUET[16], and can lead to an uncorrect evaluation of $\lim \lambda^{\epsilon}$. Examples of discrepancies between the guess and the correct limit have been exhibited by DE BUHAN[11] and TURGEMAN&col[17]. A simplified form of these examples is given in the Appendix, together with other considerations on the strength of multilayered materials. The present paper gives a variational formulation of the desired limit (Theorem 2 + corollary 2):

$$(1.8)\ \lim_{\epsilon\to 0} \lambda^{\epsilon} = \operatorname*{Inf}_{\substack{u,\mu \\ \hat{L}(u,\mu)=1}} \left\{ \int_{\Omega} j^{hom}(e(u))\, dx + \int_{\Gamma_0} j^{hom}(-u\otimes_s n)\, ds + \right.$$

$$\left. + \int_{\Gamma_1} h(x,\mu-uds) \right\}$$

$$\text{where : } \hat{L}(u,\mu) = \int_{\Omega} f_0 . u\, dx + \int_{\Gamma_1} g_0 . \mu \quad ,$$

and where $h(x,z)$ is a convex function, positively homogeneous of degree one, *which can be strictly smaller than* $j^{hom}(-z\otimes_s n(x))$.

Expressed in mathematical terms, the basic explanation of this result is that the linear form L is not lower semi-continuous in the natural functional space of definition for (1.1). There appears a relaxation term for the constraint $L(u) = 1$, and this relaxation is expressed by μ, $\hat{L}$, and h.

Expressed in physical terms, our results states that *there is a change in the behaviour of the homogenized material on loaded boundaries*. This change in behaviour is better displayed on the primal characterization of λ^{ϵ}, namely:

$$(1.9)\ \lambda^{\epsilon} = \operatorname{Sup}\ \{\ \lambda,\ \exists\ \sigma,\ \operatorname{div}(\sigma)+\lambda f_0 = 0,\ \sigma.n = \lambda g_0 \text{ on } \Gamma_0,\ \sigma(x) \in P^{\epsilon}(x) \text{ a.e. } x \text{ in } \Omega\ \}.$$

where $P^{\epsilon}(x)$ is the domain of $(j^{\epsilon})^{*}(x,.)$, hereafter called the *strength domain of the material*. The A.V.P. suggests that the limit of λ^{ϵ} could be:

$$(1.9)\ \lambda^{0} = \operatorname{Sup}\ \{\ \lambda,\ \exists\ \sigma,\ \operatorname{div}(\sigma) + \lambda f_0 = 0,\ \sigma.n = \lambda g_0 \text{ on } \Gamma_0,\ \sigma(x) \in P^{hom} \text{ a.e. } x \text{ in } \Omega\ \},$$

where P^{hom} is the domain of $(j^{hom})^{*}$.

Indeed it is proven in this paper that the limit of λ^{ϵ} is equal to λ^{hom}:

$$(1.10)\ \lambda^{hom} = \operatorname{Sup}\ \{\ \lambda,\ \exists\ \sigma,\ \operatorname{div}(\sigma)+\lambda f_0 = 0,\ \sigma.n = \lambda g_0 \text{ on } \Gamma_0,\ \sigma(x) \in P^{hom} \text{ a.e. } x \text{ in } \Omega,\ \sigma(x).n(x) \in C(x) \text{ on } \Gamma_1\ \},$$

The convex set $C(x)$, whose detailed derivation will be given in the text, denotes the strength of the material on Γ_1. It can be strictly smaller than the set $C^{hom}(x) = \{\ \sigma.n,\ \sigma \in P^{hom}\}$, indicating that the strength of the homogenized material can be strictly smaller on the loaded boundary than at any interior point of the body.

2. NOTATIONS AND ASSUMPTIONS

2.1. The Mechanical setting

A periodically non homogeneous material with limited strength occupies a domain Ω in $\mathbb{R}^N$. Throughout this paper Ω is supposed to satisfy:

(2.1) Ω is bounded and $\partial\Omega$ is C^1.

For the sake of simplicity it is assumed that $\partial\Omega$ can be shared into two compact disconnected parts Γ_0 and Γ_1.

The condition of limited strength is expressed by the fact that the stress tensor σ (symmetric N×N tensor field) belongs to a *strength domain* $P^\epsilon(x)$ at every point x in Ω. P^ϵ is assumed to be ϵY periodic:

(2.2) $\sigma(x) \in P^\epsilon(x) = P(\frac{x}{\epsilon})$ for every x in Ω.

Throughout the paper we assume that P^ϵ has the following properties:

(2.3) P^ϵ is a closed and convex subset of $\mathbb{R}_s^{N^2}$ (space of symmetric N×N tensors),

(2.4) There exists two strictly positive scalars k_0 and k_1 such that:

$$\{\sigma, |\sigma| \leq k_0\} \subset P^\epsilon(x) \subset \{\sigma, |\sigma| \leq k_1\} \text{ for every x in } \Omega.$$

(2.5) P(y) is constant on smooth subdomains of Y. The typical situation is that of a partition of Y into two subdomains Y_0 and Y_1, called the constituents, and:

$$P(y) = P^1 \text{ if } y \in Y_1, \quad P(y) = P^0 \text{ if } y \in Y_0,$$

where P^1 and P^0 are the strength domains of each constituent.

Ω is loaded by body forces λf_0 and surface forces λg_0 on the part Γ_1 of $\partial\Omega$. λ is the *load parameter*. Equilibrium of the body reads as:

(2.6) $\mathrm{div}(\sigma) + \lambda f_0 = 0$ in Ω, $\sigma.n = \lambda g_0$ on Γ_1.

Throughout the paper we shall assume the following regularity of the loading:

$$(2.7) \qquad f_0 \in \mathbb{L}^\infty(\Omega) \ , \ g_0 \in \mathbb{C}^0(\Gamma_1).$$

The *limit load*, defined statically (in terms of stress tensors), is:

$$(2.8) \quad \lambda^\epsilon = \text{Sup} \ \{ \ \lambda \mid \text{there exists } \sigma \text{ such that: } \operatorname{div}(\sigma) + \lambda f_0 = 0 \text{ in } \Omega, \ \sigma.n = \lambda g_0 \text{ on } \Gamma_1, \ \sigma(x) \in P^\epsilon(x) \text{ a.e. } x \text{ in } \Omega \ \}.$$

λ^ϵ can alternatively be determined by a dual problem TEMAM[17]:

$$(2.9) \quad \lambda^\epsilon = \operatorname*{Inf}_u \left\{ \int_\Omega j^\epsilon(e(u)) \, dx \ ; \ u = u_0 \text{ on } \Gamma_0 \ , \ L(u) = 1 \right\},$$

where :

$$(2.10) \quad j^\epsilon(x,.) = (\mathbb{I}_{P^\epsilon(x)})^* \ , \ L(u) = \int_\Omega f_0 . u \, dx + \int_{\Gamma_1} g_0 . u \, ds.$$

We shall consider in the sequel a slightly more general version of (2.9), namely:

$$(2.11) \quad \lambda^\epsilon = \operatorname*{Inf}_u \left\{ \int_\Omega j(\frac{x}{\epsilon}, e(u)) \, dx \ ; \ u = u_0 \text{ on } \Gamma_0 \ , \ L(u) = 1 \right\},$$

where u_0, and $j(y,E)$ are assumed to obey:

$$(2.12) \qquad u_0 \in \mathbb{H}^{1/2}(\Gamma_0),$$

$j(y,E)$ is convex and lower semi-continuous on $\mathbb{R}_s^{N^2}$ with respect to E, and periodic with respect to y. Moreover there exists k_0, k_1, k_2 strictly positive constants such that:

$$(2.13) \qquad k_0 \ (|E|-1) \leq j(y,E) \leq k_1 \ (|E|+1),$$

$$(2.14) \qquad j^*(y,\Sigma) \leq k_2 \text{ for every } \Sigma \text{ in } P(y) = \operatorname{dom}(j^*(y,.)).$$

Note that these assumptions imply in turn the following estimate for the singular part j_∞ of j (defined below):

$$(2.15) \qquad j_\infty(y,E) \leq j(y,E) + k_2 .$$

2.2 Functional analysis

We shall extensively use in the sequel the space of vector fields with Bounded Deformation (SUQUET[15], TEMAM[17]):

$$BD(\Omega) = \left\{ u = (u_i)_{i=1,N} \ , \ u_i \in L^1(\Omega) \ , \ \epsilon_{ij}(u) \in M^1(\Omega) \right\}$$

where $M^1(\Omega)$ stands for the space of bounded measures on Ω. Classical results assert that $BD(\Omega)$ has the following properties:

(2.16) * $BD(\Omega)$ is the dual of a Banach space and, therefore, can be endowed with a weak * topology for which bounded sets are relatively compact sets. However *$BD(\Omega)$ is not a reflexive space.*

(2.17) * There exists a trace application from $BD(\Omega)$ onto $L^1(\partial\Omega)^N$, continuous for the strong topologies of these two spaces. However *this trace application is not continuous for the weak * topologies of these spaces.*

The following space will also be useful in the sequel:

(2.18) $$LD(\Omega) = \left\{ u = (u_i)_{i=1,N} \ , \ u_i \in L^1(\Omega) \ , \ \epsilon_{ij}(u) \in L^1(\Omega) \right\}$$

In the following of the paper we shall deal with vector fields or tensorial fields, rather than with scalar fields, and we shall denote by a barred or a curved letter the corresponding functional spaces. For instance:

(2.19) $$\mathbb{L}^p(\Omega) = \left\{ u = (u_i)_{i=1,N} \ , \ u_i \in L^p(\Omega) \right\},$$

(2.20) $$\mathcal{L}^p(\Omega) = \left\{ \sigma = (\sigma_{ij})_{1 \le i,j \le N} \ , \ \sigma_{ij} = \sigma_{ji}, \ \sigma_{ij} \in L^p(\Omega) \right\}$$

2.3. Convex analysis

For a proper convex function $F : X \to \mathbb{R} \cup \{+\infty\}$ (where X is a Banach space) we define its conjugate function $F^* : X^* \to \mathbb{R} \cup \{+\infty\}$ as:

$$F^*(x^*) = \sup_{x \in X} \{ (x, x^*) - F(x) \}$$

F^* is a convex l.s.c. proper function on X^*. Moreover $\bar{F} = (F^*)^*$.

The indicator function of a nonempty, closed and convex set K in X will be denoted by $\mathbb{I}_K$:

$$\mathbb{I}_K(x) = 0 \text{ if } x \in K, \ +\infty \text{ otherwise.}$$

$(\mathbb{I}_K)^*$ is the support function of K. It is a l.s.c., convex function, positively homogeneous of degree one.

We define the recession function j_∞ of a l.s.c., convex function j by:

$$j_\infty(E) = \lim_{t\to+\infty} \frac{1}{t} j(tE) .$$

It can be easily checked that j_∞ is the support function of $\mathrm{dom}(j^*)$.

4. Γ-convergence

We shall frequently refer to the theory of Γ-convergence (DE GIORGI[12], ATTOUCH[2]). Let (X,τ) be a topological vector space, and F^ϵ a sequence of functions mapping X into $\mathbb{R}\cup\{+\infty\}$. We define $\Gamma\text{-liminf}(F^\epsilon)$ and $\Gamma\text{-limsup}(F^\epsilon)$ as follows:

$$\Gamma\text{-liminf}(F^\epsilon)(x) = \mathrm{Inf}\left\{ \mathrm{liminf}(F^\epsilon(x^\epsilon)) \; ; \; x^\epsilon \overset{\tau}{\to} x \right\}, \tag{2.21}$$

$$\Gamma\text{-limsup}(F^\epsilon)(x) = \mathrm{Inf}\left\{ \mathrm{limsup}(F^\epsilon(x^\epsilon)) \; ; \; x^\epsilon \overset{\tau}{\to} x \right\}. \tag{2.22}$$

These two Γ-limits are lower semi-continuous on (X,τ). *F^ϵ is said to be τ- Γ-convergent to F* if:

$$\Gamma\text{-liminf}(F^\epsilon) = \Gamma\text{-limsup}(F^\epsilon) . \tag{2.22}$$

It is easily checked that F^ϵ is τ-Γ-convergent to F if the two following requirements are met:

i) For every x in X, and every x^ϵ in X τ-converging to x, then

$$F(x) \le \liminf_\epsilon F^\epsilon(x^\epsilon) \tag{2.23}$$

ii) For every x in X, there exists x^ϵ in X τ-converging to x such that:

$$F(x) \ge \limsup_\epsilon F^\epsilon(x^\epsilon) \tag{2.24}$$

In the sequel we shall omit the "τ" for brevity.

We define the l.s.c. hull of any functional $F: X \to \mathbb{R}\cup\{+\infty\}$ by:

$$\bar{F} = \mathrm{Sup}\{ G, \; G(x) \le F(x) \text{ for every } x \text{ in } X, \; G \text{ is } \tau \text{ l.s.c.} \}$$

We shall sometimes refer to $\bar{F}$ as the relaxed function associated with F. An interesting property of Γ-convergence is that $\bar{F}$ is the Γ-limit of the sequence $F^\epsilon = F$.

3.HOMOGENIZATION AND RELAXATION

3.1 Preliminary result

Let us first consider the case where $\Gamma_1 = \emptyset$, and define on $\mathbb{L}^p(\Omega)$ (p will be specified later on):

(3.1) $$J^\epsilon(u) = \begin{cases} \int_\Omega j^\epsilon(e(u))\,dx & \text{if } u \in LD(\Omega), \\ +\infty \text{ otherwise.} & \end{cases},$$

(3.2) $$J^\epsilon_{u_0}(u) = J^\epsilon(u) + \mathbb{I}_{\{u = u_0 \text{ on } \partial\Omega\}}.$$

Let $j^{hom}(E)$ be the homogenized density of energy deduced from the A.V.P.:

(3.3) $$j^{hom}(E) = \operatorname*{Inf}_{w \in \mathbb{H}^1_{per}} \left\{ \frac{1}{|Y|} \int_Y j(y, E+e(w))\,dy \right\}.$$

It is readily seen that j^{hom} is convex and obeys (2.12) and (2.13). Its conjugate function reads as (BOUCHITTE[5]):

(3.4) $$(j^{hom})^*(\Sigma) = \operatorname*{Inf}_{\sigma \in S_{per}} \left\{ \frac{1}{|Y|} \int_Y j^*(y, \Sigma+\sigma(y))\,dy \right\},$$

where :

$$S_{per} = \left\{ \sigma \in \mathcal{L}^2(Y),\ \operatorname{div}(\sigma) = 0,\ \int_Y \sigma\,dy = 0,\ \sigma.n \text{ anti-periodic} \right\}.$$

We note $P^{hom} = \operatorname{dom}((j^{hom})^*)$, and we define:

(3.5) $$J^{hom}(u) = \begin{cases} \int_\Omega j^{hom}(e(u)) & \text{if } u \in BD(\Omega), \\ +\infty \text{ otherwise.} & \end{cases}$$

(3.6) $$J^{hom}_{u_0}(u) = J^{hom}(u) + \int_{\partial\Omega} j^{hom}_\infty((u_0 - u)\otimes_s n)\,ds.$$

THEOREM 1 (BOUCHITTE[5]): Let p be such that $1 \leq p \leq N/(N-1)$. Then:

i) J^{ϵ} *Γ-converges to* J^{hom} *in* $\mathbb{L}^p(\Omega)$ *weak (and strong if* $p < N/(N-1)$*)*

ii) $J^{\epsilon}_{u_0}$ *Γ-converges to* $J^{hom}_{u_0}$ *in* $\mathbb{L}^p(\Omega)$ *weak (and strong if* $p < N/(N-1)$.

The reader is referred to [5] for the proof of Theorem 1. This result has been extended to a more general form of j by DEMENGEL & TANG-QI [13].

We now turn to the problem of limit loads. Let L be the linear form:

(3.7) $$L(u) = \int_{\Omega} f_0 . u \, dx \ , \text{ where f satisfies (2.7)},$$

and consider the variational problems:

(3.8) $$\lambda^{\epsilon} = \inf_{L(u) = 1} J_0^{\epsilon}(u) \ ,$$

(3.9) $$\lambda^{hom} = \inf_{L(u) = 1} J_0^{hom}(u) \ .$$

COROLLARY 1: Under the above assumptions:

$$\lim_{\epsilon \to 0} \lambda^{\epsilon} = \lambda^{hom}.$$

Corollary 1 is a direct consequence of Theorem 1, and of the continuity of L on $\mathbb{L}^p(\Omega)$. We now turn to the more difficult general case, where a part of the boundary is loaded.

3.2 Loaded boundary: statement of the result

When Γ_1 is not empty, and g_0 does not vanish identically, a new difficulty arise, since the linear form L:

(3.10) $$L(u) = \int_{\Omega} f_0 . u \, dx + \int_{\Gamma_1} g_0 . u \, ds \ ,$$

is no more continuous on $\mathbb{L}^p(\Omega)$ or even on $BD(\Omega)$ weak* (the trace operator is not continuous from $BD(\Omega)$ weak* into $\mathbb{L}^1(\Gamma_1)$). The constraint { L(u) =1 } is therefore not closed for the natural topology for which minimizing sequences of the variational problem

(3.8) contain a converging subsequence.

To overcome this difficulty, we consider separately the contribution of the displacement rate u in the inside of Ω and its contribution on the boundary Γ_1 which is denoted by μ. Since μ is in $\mathbb{L}^1(\Gamma_1)$ as soon as u is in BD(Ω), the most convenient functional space for μ is $\mathbb{M}^1(\Gamma_1)$, where bounded sequences in $\mathbb{L}^1(\Gamma_1)$ contain weakly* convergent subsequences. Let p be such that $1< p < N/(N-1)$ and define on $\mathbb{X}=\mathbb{L}^p(\Omega)\times\mathbb{M}^1(\Gamma_1)$ a sequence of functionals Φ^ϵ by:

$$(3.11)\quad \Phi^\epsilon(u,\mu) = \begin{cases} \int_\Omega j^\epsilon(x,e(u))\,dx \quad , \text{ if } u \in LD(\Omega) \ , \quad u|_{\Gamma_0} = u_0 \ , \\ \qquad\qquad\qquad \text{and } \mu = u\,ds \text{ on } \Gamma_1 . \\ +\infty \quad \text{otherwise} \end{cases}$$

and a linear form, continuous on $\mathbb{X}$, by:

$$(3.12)\qquad \hat{L}(u,\mu) = \int_\Omega f_0 . u\,dx + \int_{\Gamma_1} g_0\,d\mu \ .$$

Note that:

$$(3.13)\ \lambda^\epsilon = \operatorname*{Inf}_{u,\mu}\ \{ \Phi^\epsilon(u,\mu) \ , \ \hat{L}(u,\mu) = 1\} = \operatorname*{Inf}_{u}\ \{ J^\epsilon(u) \ , \ L(u) = 1\}.$$

Since $\hat{L}$ is continuous (while L was not), it is sufficient, in order to pass to the limit in (3.13), to determine the Γ limit of Φ^ϵ on $\mathbb{X}$ endowed with the strong topology of $\mathbb{L}^p(\Omega)$ and the weak* topology of $\mathbb{M}^1(\Gamma_1)$ (note that due to the equi-coercivity of Φ^ϵ, and to the compact embedding of BD(Ω) into $\mathbb{L}^p(\Omega)$, it would be equivalent to search for the Γ-limit of Φ^ϵ into BD(Ω) $\times$ $\mathbb{M}^1(\Gamma_1)$ endowed with the weak* topology of each space).

For this purpose, consider:

$$S(\Omega) = \left\{ \sigma \in \mathcal{L}^\infty(\Omega), \ \operatorname{div}(\sigma) \in \mathbb{L}^{p'}(\Omega) \ , \ \sigma.n \in \mathbb{C}^0(\Gamma_1) \right\} ,$$

and for f in $\mathbb{L}^{p'}(\Omega)$ and g in $\mathbb{C}^0(\Gamma_1)$:

$$S(f,g) = \{ \sigma \in S(\Omega), \ \operatorname{div}(\sigma) + f = 0, \ \sigma.n = g \text{ on } \Gamma_1 \ \}$$

S(Ω) is endowed with the following topology τ:

$$(3.14)\quad \sigma^\epsilon \overset{\tau}{\rightarrow} \sigma \quad \text{iff} \quad \begin{cases} \sigma^\epsilon \rightarrow \sigma \text{ in } \mathcal{L}^\infty(\Omega) \text{ weak*} \\ \operatorname{div}(\sigma^\epsilon) \rightarrow \operatorname{div}(\sigma) \text{ in } \mathbb{L}^{p'}(\Omega) \text{ weak *} \\ \sigma^\epsilon.n|_{\Gamma_1} \rightarrow \sigma.n|_{\Gamma_1} \text{ in } \mathbb{C}^0(\Gamma_1) \text{ strong} \end{cases}$$

Moreover let us define:

$$\mathbb{K}^{\epsilon} = \{ \sigma \in S(\Omega),\ \sigma(x) \in P^{\epsilon}(x) \text{ a.e. } x \text{ in } \Omega\}$$

When ϵ goes to 0, $\mathbb{K}^{\epsilon}$ admits a limsup, denoted by $\mathbb{K}^{s}$, and a liminf, denoted by $\mathbb{K}^{i}$, in Kuratowski's sense for the topology τ. *We further assume that there exists a closed subset $\mathbb{K}$ in $S(\Omega)$ such that:*

(H) $\mathbb{K}^{\epsilon}$ converges to $\mathbb{K}$ in Kuratowski's sense for the topology τ on $S(\Omega)$.

Then we define for every x in Γ_1:

$$C(x) = \text{Closure}\{ \sigma.n(x),\ \sigma \in \mathbb{K} \}.$$

C(x) is a l.s.c. multi-application with convex and closed values.

We are now in a position to state our main result:

<u>THEOREM 2:</u> *The Γ-limit of Φ^{ϵ} in $\mathbb{X}$ is:*

$$\Phi^{hom}(u,\mu) = \int_{\Omega} j^{hom}(e(u)) + \int_{\Gamma_0} j^{hom}_{\infty}((u_0-u)\otimes_s n)\, ds + \int_{\Gamma_1} h(x,\mu-uds),$$

where $h(x,z)$ is the l.s.c., convex, positively homogeneous of degree 1 function defined as:

$$h(x,z) = \mathbb{I}^{*}_{C(x)}(z).$$

Consider now the sequence λ^{ϵ} of infima (3.13) (where $u_0 = 0$), and set:

$$\text{(3.15)} \qquad \lambda^{hom} = \operatorname*{Inf}_{u,\mu \in \mathbb{X}} \{ \Phi^{hom}(u,\mu)\ ,\ \hat{L}(u,\mu) = 1 \}.$$

<u>COROLLARY 2:</u> *Under the above assumptions:*

$$\lim_{\epsilon \to 0} (\lambda^{\epsilon}) = \lambda^{hom}.$$

Corollary 2 is a direct consequence of Theorem 2 and of the

continuity of $\hat{L}$ on $\mathfrak{X}$.

Remark 1: It can be checked (cf [7]) that:

$$\lambda^{hom} = \mathrm{Inf}(\ \lambda^0,\ \Lambda\ ),$$

where:

$$\lambda^0 = \underset{L(u)\,=\,1}{\mathrm{Inf}} \left\{ J_0^{hom}(u) \right\}, \text{ and}$$

$$\Lambda = \mathrm{Inf}\left\{ \int_{\Gamma_1} h(x,\mu)\ ,\ \int_{\Gamma_1} g\mu = 1\ ,\ \mu \in \mathbb{M}^1(\Gamma_1) \right\}$$

$$= \mathrm{Sup}\ \{\ \lambda\ ,\ \lambda g(x) \in C(x) \text{ for every } x \text{ in } \Gamma_1\ \}$$

We begin the proof of Theorem 2 with a preliminary result.

PROPOSITION 1: Let σ^ϵ be a sequence of elements of $S(\Omega)$ converging to σ in $S(\Omega)$ for the above described topology τ. Then:

$$\text{(3.16)} \qquad \liminf_{\epsilon\to 0} \int_\Omega (j^\epsilon)^*(\sigma^\epsilon)\, dx \geq \int_\Omega (j^{hom})^*(\sigma)\, dx\ .$$

Moreover if σ^ϵ belongs to $\mathbb{K}^\epsilon$, then σ belongs to $\mathbb{K}^{hom}$.

Proof of Proposition 1:

Let $(\Omega_i)_{i\in I}$ be a finite family of disconnected, Lipschitzian open subsets of Ω, such that

$$\mathrm{mes}(\Omega - \bigcup_{i\in I} \Omega_i) = 0.$$

Let:

$$\text{(3.17)} \qquad z = \sum_{i\in I} z_i\ \chi_{\Omega_i}(x)\ ,$$

where $z_i \in \mathbb{R}_s^{N^2}$, and χ_A is the characteristic function of A. Application of Theorem 1 to each open set Ω_i and to the sequence of functions $j^\epsilon(x, z_i + .)$, yields the existence of a sequence u_ϵ^i in $BD(\Omega_i)$ such that:

i) $u_\epsilon^i |_{\partial\Omega_i} = 0$, $\lim_{\epsilon\to 0}(u_\epsilon^i) = 0$ in $\mathbb{L}^{p'}(\Omega_i)$,

ii) $\lim_{\epsilon\to 0} \int_{\Omega_i} j^\epsilon(z_i + e(u_\epsilon^i))\, dx = \int_{\Omega_i} j^{hom}(z_i)\, dx.$

Let $u^\epsilon = \sum_{i\in I} u_\epsilon^i \chi_{\Omega_i}$. Then:

i) $u^\epsilon |_{\partial\Omega} = 0$, $\lim_{\epsilon\to 0} (u^\epsilon) = 0$ in $\mathbb{L}^{p'}(\Omega)$,

ii) $\lim_{\epsilon\to 0} \int_\Omega j^\epsilon(Z+e(u^\epsilon))\, dx = \int_\Omega j^{hom}(Z)\, dx.$

Now let σ^ϵ be a sequence in $S(\Omega)$ τ-converging to σ. Fenchel's inequality yields:

$$\int_\Omega (j^\epsilon)^*(\sigma^\epsilon)\, dx \geq \int_\Omega \sigma^\epsilon : (Z+e(u^\epsilon))\, dx - \int_\Omega j^\epsilon(Z+e(u^\epsilon))\, dx.$$

But:

$$\lim_{\epsilon\to 0} \int_\Omega \sigma^\epsilon : e(u^\epsilon)\, dx = - \lim_{\epsilon\to 0} \int_\Omega \mathrm{div}(\sigma^\epsilon).u^\epsilon\, dx = 0,$$

therefore:

$$\liminf_{\epsilon\to 0} \int_\Omega (j^\epsilon)^*(\sigma^\epsilon)\, dx \geq \int_\Omega (\sigma : Z - j^{hom}(Z))\, dx. \tag{3.18}$$

(3.18) is valid for every piecewise constant Z in the form (3.17), and can be extended by density to every Z in $\mathcal{L}^1(\Omega)$. We take the supremum of the right hand side of (3.18) on Z in $\mathcal{L}^1(\Omega)$, and note that by Rockafellar's theorem:

$$\sup_{Z\in\mathcal{L}^1(\Omega)} \left(\int_\Omega (\sigma : Z - j^{hom}(Z))\, dx \right) = \int_\Omega (j^{hom})^*(\sigma)\, dx . \tag{3.19}$$

(3.16) is now a direct consequence of (3.18) and (3.19).

If moreover σ^ϵ belongs to $\mathbb{K}^\epsilon$, then the left hand side of (3.16) is finite. The same conclusion holds for the right hand side, and $\sigma(x)$ is for a.e. x in the domain of $(j^{hom})^*$, i.e. in P^{hom}. Q.E.D.

Remark 2: Proposition 1 can be strengthened in the following way . Let:

$$G^{\epsilon}(\sigma) = \int_{\Omega} (j^{\epsilon})^{*}(\sigma)\, dx \ , \quad G^{hom}(\sigma) = \int_{\Omega} (j^{hom})^{*}(\sigma)\, dx.$$

Then G^{ϵ} Γ-converges to G^{hom} in $S(\Omega)$ endowed with the topology τ. This stronger result will not be useful in the sequel.

3.3 Proof of Theorem 2. First step: liminf Φ^{ϵ}

In a first step we show that:

$$\text{(3.20)} \qquad \liminf \Phi^{\epsilon} \geq \Phi^{hom} \ .$$

For the sake of simplicity we only consider the case $\Gamma_1 = \partial\Omega$.
Let $(u^{\epsilon},\mu^{\epsilon})$ be a sequence in $\mathbb{X}$ converging to (u,μ) for the topology $\mathbb{L}^p(\Omega)$ strong $\times$ $\mathbb{M}^1(\Gamma_1)$ weak $*$, and such that:

i) $\Phi^{\epsilon}(u^{\epsilon},\mu^{\epsilon}) \leq C$,
(which implies that $u^{\epsilon} \in LD(\Omega)$ and $\mu^{\epsilon} = u^{\epsilon}ds$)

ii) $\lim\limits_{\epsilon\to 0} (u^{\epsilon}) = u$ in $\mathbb{L}^p(\Omega)$ strong,
$\lim\limits_{\epsilon\to 0} (u^{\epsilon}ds) = \mu$ in $\mathbb{M}^1(\partial\Omega)^N$ weak*,

Let σ be an element of $\mathbb{K}$, and σ^{ϵ} a sequence of elements of $\mathbb{K}^{\epsilon}$ τ-converging to σ. Consider, after ANZELLOTTI [1], the measure $\lambda^{\epsilon} = \sigma^{\epsilon}:e(u^{\epsilon})$ defined on Ω by:

$$\text{(3.21)} \quad \langle\lambda^{\epsilon},\varphi\rangle = - \int_{\Omega} \text{div}(\sigma^{\epsilon}).u^{\epsilon}\varphi\, dx - \int_{\Omega} \sigma^{\epsilon}:(u^{\epsilon}\otimes\text{grad}(\varphi))\, dx.$$

The topologies for which σ^{ϵ} and u^{ϵ} are converging sequences allow us to pass to the limit in the right hand side of (3.21):

$$\lim_{\epsilon\to 0} \langle\lambda^{\epsilon},\varphi\rangle = \langle\lambda,\varphi\rangle = -\int_{\Omega} \text{div}(\sigma).u\varphi\, dx - \int_{\Omega} \sigma:(u\otimes\text{grad}(\varphi))\, dx.$$

Therefore λ^{ϵ} converges in $M^1(\Omega)$ weak* to $\lambda = \sigma:e(u)$. Define:

$$\Omega_{\alpha} = \{ x \in \Omega, \ \text{dist}(x,\partial\Omega) > \alpha \}$$

Then a classical argument asserts that, for α outside a countable set, the following convergence holds:

$$\text{(3.22)} \qquad \lim_{\epsilon\to 0} \lambda^{\epsilon}(\Omega_{\alpha}) = \lambda(\Omega_{\alpha}) \ .$$

Consider such an α as fixed for a moment. Then:

$$(3.23)\quad \int_{\Omega} j^{\epsilon}(e(u^{\epsilon}))\,dx = \int_{\Omega_\alpha} j^{\epsilon}(e(u^{\epsilon}))\,dx + \int_{\Omega-\Omega_\alpha} j^{\epsilon}(e(u^{\epsilon}))\,dx,$$

and by Theorem 1:

$$(3.24)\quad \lim_{\epsilon\to 0} \int_{\Omega_\alpha} j^{\epsilon}(e(u^{\epsilon}))\,dx \geq \int_{\Omega_\alpha} j^{hom}(e(u))\,dx.$$

Next we note that:

$$\int_{\Omega-\Omega_\alpha} j^{\epsilon}(e(u^{\epsilon}))\,dx \geq \int_{\Omega-\Omega_\alpha} j^{\epsilon}_{\infty}(e(u^{\epsilon}))\,dx - k_2\,|\Omega-\Omega_\alpha|\ ,$$

$$(3.25)\qquad \geq \int_{\Omega-\Omega_\alpha} \sigma^{\epsilon}:e(u^{\epsilon})\,dx - k_2\,|\Omega-\Omega_\alpha|\ .$$

But:

$$\int_{\Omega} \sigma^{\epsilon}:e(u^{\epsilon})\,dx = \int_{\partial\Omega} \sigma^{\epsilon}.n.u^{\epsilon}\,ds - \int_{\Omega} \mathrm{div}(\sigma^{\epsilon}).u^{\epsilon}\,dx.$$

Considering the convergence of each term under the integrals in the right hand side of this equality, we obtain on one hand:

$$\lim_{\epsilon\to 0} \int_{\Omega} \sigma^{\epsilon}:e(u^{\epsilon})\,dx = \int_{\partial\Omega} \sigma.n\,d\mu - \int_{\Omega} \mathrm{div}(\sigma).u\,dx\ .$$

On the other hand, by virtue of (3.22), we have:

$$\lim_{\epsilon\to 0} \int_{\Omega_\alpha} \sigma^{\epsilon}:e(u^{\epsilon})\,dx = \int_{\Omega_\alpha} \sigma:e(u)\,dx\ .$$

Therefore:

$$\lim_{\epsilon\to 0} \int_{\Omega-\Omega_\alpha} \sigma^{\epsilon}:e(u^{\epsilon})\,dx = \int_{\partial\Omega} \sigma.n\,d\mu - \int_{\Omega} \mathrm{div}(\sigma).u\,dx - \int_{\Omega_\alpha} \sigma:e(u)\,dx$$

Coming back to (3.23)(3.24) and (3.25) we obtain:

$$\liminf_{\epsilon\to 0} \Phi^{\epsilon}(u^{\epsilon},\mu^{\epsilon}) \geq \int_{\Omega_\alpha} j^{hom}(e(u)) + \int_{\partial\Omega} \sigma.n\,d\mu - \int_{\Omega} \mathrm{div}(\sigma).u\,dx$$
$$- \int_{\Omega_\alpha} \sigma:e(u)\,dx - k_2\,|\Omega-\Omega_\alpha|\ .$$

Now letting α go to 0, and after due use of Green's formula, we obtain:

$$(3.26) \qquad \liminf_{\epsilon \to 0} \Phi^{\epsilon}(u^{\epsilon},\mu^{\epsilon}) \geq \int_{\Omega} j^{hom}(e(u)) + \int_{\partial\Omega} \sigma.n\ (d\mu - uds).$$

(3.26) is valid for every σ in $\mathbb{K}$, and we take the supremum of its right hand side with respect to such σ's. To compute this supremum we claim that:

$$(3.27) \qquad \operatorname*{Sup}_{\sigma \in \mathbb{K}} \int_{\partial\Omega} \sigma.n.(d\mu - uds) = \int_{\partial\Omega} \operatorname*{Sup}_{z \in C(x)} (z.(d\mu - uds))$$

$$= \int_{\partial\Omega} h(x,\mu - uds).$$

To prove this claim it is sufficient to prove that the set $\{ \sigma.n,\ \sigma \in \mathbb{K} \}$ is stable under Lipschitzian partition of unity (see BOUCHITTE & VALADIER [8]). This is the object of Lemma 1 below. (3.27) (3.26) complete the proof of (3.20).

<u>*LEMMA 1*</u>: *Define* $\mathbb{K}_1 = \{ \sigma.n,\ \sigma \in \mathbb{K} \}$. *Let* $(\varphi_i)_{i \in I}$ *be a finite family of elements in* $\mathbb{K}_1$, *and* $(\alpha_i)_{i \in I}$ *a Lipschitzian partition of unity on* $\partial\Omega$:

$$\alpha_i \in Lip(\partial\Omega,[0,1]) \ ,\ \sum_{i \in I} \alpha_i = 1.$$

Then: $$\varphi = \sum_{i \in I} \alpha_i\ \varphi_i \textit{ belongs to } \mathbb{K}_1 .$$

<u>Proof of Lemma 1:</u>

By definition of $\mathbb{K}_1$ there exists a family $(\sigma_i)_{i \in I}$ of elements in $\mathbb{K}$ such that $\varphi_i = \sigma_i.n$ on $\partial\Omega$. According to BOUCHITTE & VALADIER [8], the partition of unity $(\alpha_i)_{i \in I}$ on $\partial\Omega$ can be extended into a Lipschitzian partition of unity $(\beta_i)_{i \in I}$ on $\bar{\Omega}$.

Then $\sigma = \sum_{i \in I} \beta_i\ \sigma_i$ satisfies :

$$\varphi = \sigma.n \text{ on } \partial\Omega.$$

To complete the proof of Lemma 1 it remains to prove that σ belongs to $\mathbb{K}$. By definition of $\mathbb{K}$, each σ_i can be approached in the topology (3.14) by a sequence σ_i^{ϵ} of elements in $\mathbb{K}^{\epsilon}$. Define $\sigma^{\epsilon} = \sum_{i \in I} \beta_i\ \sigma_i^{\epsilon}$. It is easily checked that σ^{ϵ} belongs to $\mathbb{K}^{\epsilon}$ (convexity of P^{ϵ}, and Lipschitz regularity of β_i), and moreover that σ^{ϵ} converges to σ in the topology (3.14). Therefore σ belongs to $\mathbb{K}$, and Lemma 1 is proved. Q.E.D

3.4 Proof of Theorem 2. Second step: limsup Φ^{ϵ}

In a second step we show that:

(3.30) $$\Gamma\text{-}\limsup_{\epsilon\to 0} \Phi^{\epsilon} \leq \Phi^{hom}.$$

LEMMA 2: In the duality between $\mathbb{X}$ *and* $\mathbb{L}^{p'}(\Omega)\times\mathbb{C}^0(\Gamma_1)$, *the conjugate functions of* Φ^{ϵ} *and* Φ^{hom} *read as:*

(3.31) $$(\Phi^{\epsilon})^*(f,\varphi) = \operatorname*{Inf}_{\sigma\in S(f,\varphi)} \left\{ \int_{\Omega} (j^{\epsilon})^*(\sigma)\, dx - \int_{\Gamma_0} \sigma.n.u_0\, ds \right\}$$

(3.32) $$(\Phi^{hom})^*(f,\varphi) = \begin{cases} \displaystyle\operatorname*{Sup}_{\sigma\in S(f,\varphi)} \int_{\Omega} (j^{hom})^*(\sigma)\, dx - \int_{\Gamma_0} \sigma.n.u_0\, ds \\ \quad \text{if } \varphi(x) \in C(x) \text{ for every } x \text{ in } \Gamma_1, \\ \quad +\infty \text{ otherwise} \end{cases}$$

Let us first prove how (3.30) can be deduced from Lemma 2. For this purpose it is sufficient (AZE [3]) to establish the following inequality:

(3.33) $$\Gamma\text{-}\liminf(\Phi^{\epsilon})^* \geq (\Phi^{hom})^*.$$

Let $(f^{\epsilon}, \varphi^{\epsilon})$ be a converging sequence in $\mathbb{L}^{p'}(\Omega)$ weak$\times\mathbb{C}^0(\Gamma_1)$ strong, with limit (f,φ), and such that :

$$\liminf\, (\Phi^{\epsilon})^*(f^{\epsilon},\varphi^{\epsilon}) < +\infty .$$

According to Lemma 2, there exists a sequence σ^{ϵ} in $S(\Omega)$ such that:

i) $\sigma^{\epsilon} \in S(f^{\epsilon},\varphi^{\epsilon})$, thus $\sigma^{\epsilon} \in \mathbb{K}^{\epsilon}$,

(3.34) ii) $$(\Phi^{\epsilon})^*(f^{\epsilon},\varphi^{\epsilon}) \geq \int_{\Omega} (j^{\epsilon})^*(\sigma^{\epsilon})\, dx - \int_{\Gamma_0} \sigma^{\epsilon}.n.u_0\, ds - \epsilon.$$

It can be readily seen that the growth condition (2.13) together with (3.34) implies that σ^{ϵ} is bounded in $\mathcal{L}^{\infty}(\Omega)$. Since it belongs to $S(f^{\epsilon},\varphi^{\epsilon}) \cap \mathbb{K}^{\epsilon}$ it contains a τ-converging subsequence, the limit of which, noted σ, is in $S(f,\varphi) \cap \mathbb{K}$. Moreover :

$$(3.35) \qquad \lim_{\epsilon \to 0} \int_{\Gamma_0} \sigma^{\epsilon}.n.u_0 \; ds = \int_{\Gamma_0} \sigma.n.u_0 \; ds \; .$$

According to Proposition 1:

$$(3.36) \qquad \liminf_{\epsilon \to 0} \int_{\Omega} (j^{\epsilon})^{*}(\sigma^{\epsilon}) \; dx \geq \int_{\Omega} (j^{hom})^{*}(\sigma) \; dx \; .$$

Coming back to (3.34) with the help of (3.35)(3.36), we obtain:

$$(3.37) \qquad \liminf_{\epsilon \to 0} (\Phi^{\epsilon})^{*}(f^{\epsilon},\varphi^{\epsilon}) \geq (\Phi^{hom})^{*}(f,\varphi),$$

which is exactly the desired statement (3.30). The proof of Theorem 2 is complete, provided that we prove Lemma 2.

Proof of Lemma 2:

For (f,φ) in $\mathbb{L}^{p'}(\Omega)\times\mathbb{C}^0(\Gamma_1)$ we compute:

$$(\Phi^{\epsilon})^{*}(f,\varphi) = \sup_{u,\mu \in X} \left(\int_{\Omega} f.u \; dx + \int_{\Gamma_1} \varphi.d\mu - \Phi^{\epsilon}(u,\mu) \right)$$

$$(3.38) \qquad = - \inf_{u} \left(\int_{\Omega} j^{\epsilon}(e(u)) \; dx - \int_{\Omega} f.u \; dx - \int_{\Gamma_1} \varphi.u \; ds \right)$$

where in (3.38) $u \in LD(\Omega)$ and $u = u_0$ on Γ_0. The computation of the dual problem associated with (3.38) is a routine exercise in Convex Analysis (see TEMAM[17]), and yields (3.31).

The derivation of (3.22) contains a difficulty, since Φ^{hom} is a priori defined on the non reflexive space $BD(\Omega)\times\mathbb{M}^1(\Gamma_1)$, for which application of Convex Analysis is not straightforward. Therefore we consider in a first step the following function:

$$(3.39) \qquad \Psi(u,\mu) = \begin{cases} \Phi^{hom}(u,\mu) \text{ if } u \in LD(\Omega) \; , \; \mu \in \mathbb{L}^1(\Gamma_1), \\ +\infty \text{ otherwise.} \end{cases}$$

We claim that Ψ and Φ^{hom} have the same dual functions, or in other words that they have same l.s.c. regularized functions in $\mathbb{L}^p(\Omega)\times$ $\mathbb{M}^1(\Gamma_1)$. The following set of inequalities is straightforward:

$$\Psi \geq \Phi^{hom} \Rightarrow \bar{\Psi} \geq \bar{\Phi}^{hom}.$$

In order to prove the reverse inequality ($\bar{\Psi} \leq \bar{\Phi}^{hom}$), consider, for every (u,μ) in $BD(\Omega)\times\mathbb{M}^1(\Gamma_1)$, a sequence $(u^\epsilon, \theta^\epsilon)$ in $LD(\Omega)\times\mathbb{L}^1(\Omega)$ such that:

i) $\lim_{\epsilon\to 0} u^\epsilon = u$ in $\mathbb{L}^p(\Omega)$, $u^\epsilon = u$ on $\partial\Omega$, and

$$\lim_{\epsilon\to 0}\int_\Omega j^{hom}(e(u^\epsilon))\,dx = \int_\Omega j^{hom}(e(u)) .$$

Existence of such a sequence is ensured by TEMAM[17] (see also DAL MASO [10]).

ii) $\lim_{\epsilon\to 0} \theta^\epsilon = \mu$ in $\mathbb{M}^1(\Gamma_1)$ weak*, and

$$\lim_{\epsilon\to 0}\int_{\Gamma_1} h(x,\theta^\epsilon - u)\,ds = \int_{\Gamma_1} h(x,\mu - uds).$$

Existence of such a sequence results from BOUCHITTE & VALADIER [9] since $h(x,.) = \mathbb{I}^*_{C(x)}(.)$ where C is a l.s.c. multi-application with closed convex values. For this sequence we obtain:

$$\lim_{\epsilon\to 0}\Psi(u^\epsilon,\theta^\epsilon) = \Phi^{hom}(u,\mu) ,$$

i.e. $\bar{\Psi} \le \Phi^{hom}$ and consequently $\bar{\Psi} \le \bar{\Phi}^{hom}$. This inequality completes the proof of the fact that Ψ and Φ^{hom} have the same dual function.

We now proceed to the computation of Ψ^*. For (f,φ) in $\mathbb{L}^{p'}(\Omega)\times \mathbb{C}^0(\Gamma_1)$, $\Psi^*(f,\varphi)$ is defined as:

$$\Psi^*(f,\varphi) = \operatorname*{Sup}_{u,\theta}\left\{\int_\Omega (f.u - j^{hom}(e(u)))\,dx + \int_{\Gamma_1}(\theta.\varphi - h(x,\theta-u))\,ds\right\}$$

where the Sup is taken over $(u,\theta)\in LD(\Omega)\times\mathbb{L}^1(\Gamma_1)$, $u = u_0$ on Γ_0. For fixed u the Supremum in θ is computed by means of Rockafellar's theorem:

$$\operatorname*{Sup}_{\theta\in\mathbb{L}^1(\Gamma_1)}\left\{\int_{\Gamma_1}(\theta.\varphi - h(x,\theta-u))\,ds\right\} = \int_{\Gamma_1}(\theta.u + h^*(x,\varphi(x)))\,ds =$$

$$= \begin{cases} \displaystyle\int_{\Gamma_1}\theta.u\,ds \text{ if } \varphi(x)\in C(x) \text{ for every } x \text{ in } \Gamma_1, \\ +\infty \text{ otherwise} \end{cases}$$

Therefore:

$$(3.40)\quad \Psi^*(f,\varphi) = \begin{cases} -\operatorname*{Inf}_u \left\{ \int_\Omega (j^{hom}(e(u)) - f.u)\, dx - \int_{\Gamma_1} \varphi.u\, ds, \right\} \\ \qquad \text{if } \varphi(x) \in C(x) \text{ for every } x \text{ in } \Gamma_1, \\ \qquad +\infty \text{ otherwise.} \end{cases}$$

where the infimum in (3.40) is taken over $u \in LD(\Omega)$, $u = u_0$ on Γ_0. We can now perform a routine computation using Convex Analysis, to evaluate the infimum in (3.40) as the supremum in (3.32). Q.E.D

3.5 Comments about the assumption (H)

For practical use, we need to check assumption (H), and to determine more explicitely the convex set $C(x)$ or its support function $h(x,z)$: this is a difficult problem. However the determination of another set $\hat{C}(x)$ gives an useful estimate on $C(x)$. In several cases, the limit in Kuratowski's sense in $\mathbb{C}^0(\Gamma_1)$ strong of the sets :

$$\mathbb{K}_1^\epsilon = \left\{ \sigma.n|_{\Gamma_1},\ \sigma \in \mathbb{K}^\epsilon \right\},$$

can be more easily determined. Indeed it can been proved (BOUCHITTE [6]) that, if $\mathbb{K}_1^\epsilon$ converges in Kuratowski's sense in $\mathbb{C}^0(\Gamma_1)$ to $\mathbb{K}_1$, then $\mathbb{K}_1$ reads as:

$$\mathbb{K}_1 = \left\{ \varphi \in C^0(\Gamma_1)^N,\ \varphi(x) \in \hat{C}(x) \right\},$$

where $\hat{C}$ is a l.s.c. multi-application, with closed convex values.

In the case of two constituents, with strength domains P_0 and P_1, an explicit formula for $\hat{C}(x)$ can be derived. Specifically, let us define:

$$\Omega_1^\epsilon = \left\{ x \in \Omega,\ P^\epsilon(x) = P_1 \right\},\ A^\epsilon = \Gamma_1 \cap \bar{\Omega}_1^\epsilon .$$

We assume that:

$$(3.41)\quad H^{N-1}(\partial A^\epsilon) = 0, \quad Int(A^\epsilon) \to A,\ Int(\Gamma_1 - A^\epsilon) \to B .$$

Then ([6]) A and B are closed and:

$$(3.42)\quad \hat{C}(x) = \begin{cases} P_1\, n(x) \text{ if } x \in A-B, \\ P_0\, n(x) \text{ if } x \in B-A, \\ P_1\, n(x) \cap P_0\, n(x) \text{ if } x \in A \cap B. \end{cases}$$

More generally let us define:

$$\mathbb{K}^{hom} = \{ \sigma \in S(\Omega) \ , \ \sigma(x) \in P^{hom} \text{ a.e. } x \text{ in } \Omega\} \ ,$$

$$C^{hom}(x) = \text{closure } \{ \sigma.n(x), \ \sigma \in \mathbb{K}^{hom}\} \ .$$

Under the hypothesis (2.1) assuming that $\partial\Omega$ is C^1, it can be checked that $C^{hom}(x) = P^{hom}.n(x)$ (see lemma 3 below). The following proposition relates $C(x)$, $\hat{C}(x)$ and $C^{hom}(x)$.

PROPOSITION 2: Under the above assumptions:

$$i) \ (P_1 \cap P_0) \ n(x) \subset C(x) \subset \hat{C}(x) \cap C^{hom}(x),$$

$$ii) \ h(x,z) \leq j_\infty^{hom}(n(x) \otimes_s z) \ .$$

The inclusion in i) can be strict as well as the inequality in ii).

Remark 3: Note that a more specific result can be established if:

$$(3.43) \qquad P_0 \subset P_1 .$$

Indeed, in this case :

$$(P_1 \cap P_0)n(x) = P_0 n(x) = P_1 n(x) \cap P_0 n(x) \ ,$$

and we deduce from (3.42) and Proposition 2 that:

$$(3.44) \qquad C(x) = P_0 n(x) \text{ for every } x \text{ in } B.$$

On B the strength of the boundary is ruled by the strength of the weakest material. It is proven in the appendix that, in case of a layered composite material satisfying (3.43), this conclusion holds true for the entire boundary.

Proof of Proposition 2 :

We begin with a preliminary result:

LEMMA 3: Under assumption (2.1) the following equality holds:

$$C^{hom}(x) = P^{hom}.n(x) \text{ for every } x \text{ in } \Gamma_1 .$$

Proof of Lemma 3: First note that the inclusion

$$P^{hom}.n(x) \subset C^{hom}(x)$$

is immediate: if Z is in $P^{hom}.n(x)$ for one x in Γ_1, there exists Σ in P^{hom} such that $Z = \Sigma.n(x)$. The constant field $\sigma(x) = \Sigma$ is obviously in $\mathbb{K}^{hom}$, and by definition of $C^{hom}(x)$, $\Sigma.n(x)$ belongs to it. In order to prove the reverse inclusion, we note, after ANZELLOTTI, that for σ in $\mathcal{L}^{\infty}(\Omega)$ and $\mathrm{div}(\sigma)$ in $\mathbb{L}^{p'}(\Omega)$, and provided that $\partial\Omega$ is C^1, we have the following characterization of $\sigma.n$ on $\partial\Omega$: at every Lebesgue point of $\sigma.n$

$$\sigma.n(x) = \lim_{\rho\to 0^+} \lim_{r\to 0^+} \left\{ \frac{1}{|Q_{r,\rho}(x)|} \int_{Q_{r,\rho}(x)} \sigma.n(y)\, dy \right\}$$

where $Q_{r,\rho}(x) = \{ y - tn(x),\ |y-x| < \rho,\ 0 < t < r \}$. Therefore $\sigma.n(x)$ belongs to $P^{hom}.n(x)$ a.e. x in $\partial\Omega$, as soon as $\sigma(y)$ belongs to P^{hom} a.e. y in Ω. If moreover σ belongs to $\mathbb{K}^{hom}$, the continuity of both $\sigma.n(x)$ and of the multi-application $P^{hom}.n(x)$ imply that $\sigma.n(x)$ is in $P^{hom}.n(x)$ for every x in Γ_1. Therefore $P^{hom}.n(x)$ contains $\{ \sigma.n(x),\ \sigma \in \mathbb{K}^{hom}\}$, and since it is closed, it contains the closure of this set, which is exactly $C^{hom}(x)$. Q.E.D.

Coming back to the proof of proposition 2, we note that the first inclusion in i) is straightforward. Let us prove the second one. Let Z be an element of C(x), and $Z' = \lambda Z$ with $0 < \lambda < 1$. There exists φ in $\mathbb{C}^0(\Gamma_1)$ such that $\varphi(x) = Z'$, and σ in $\mathbb{K}$ such that $\sigma.n = \varphi$ on Γ_1. Let σ^{ϵ} be a sequence of elements in $\mathbb{K}^{\epsilon}$ converging to σ for the topology (3.14). Then, according to Proposition 1:

$$\sigma(x) \in P^{hom} \text{ a.e. } x \text{ in } \Omega,$$

and therefore $\varphi(x) = \sigma.n(x) \in P^{hom}n(x)$ for every x in Γ_1. Moreover:

$$\lim_{\epsilon\to 0} \sigma^{\epsilon}.n|_{\Gamma_1} = \sigma.n|_{\Gamma_1} \text{ in } \mathbb{C}^0(\Gamma_1),$$

and therefore $\varphi(x) = \sigma.n(x) \in \hat{C}(x)$ for every x in Γ_1, and more specifically that Z' belongs to $\hat{C}(x)$. The conclusion is extended to Z by letting λ tend to 1. This completes the proof of the second inclusion in i). ii) is a direct consequence of this second inclusion.

REFERENCES

[1] ANZELLOTTI M.: "On the existence of the rates of stress and displacement for Prandtl-Reuss Plasticity". *Quart. Appl. Math.*, 46, 1983, p 181-208.

[2] ATTOUCH H.: *Variational convergence for functions and operators*. Pitman. London. 1984.

[3] AZE D.: "Convergence des variables duales dans des problèmes de transmission à travers des couches minces par des méthodes d'épi-convergence". *Ricerce di Matematica*, 35, 1986, p 125-159.

[4] BENSSOUSSAN A., LIONS J.L., PAPANICOLAOU G.: *Asymptotic Analysis for Periodic Structures*. North Holland. Amsterdam. 1978.

[5] BOUCHITTE G.: "Convergence et relaxation de fonctionnelles du calcul des variations à croissance linéaire. Application à l'homogénéisation en Plasticité.". *Ann. Fac. Sc. Toulouse*, 8, 1986-1987, p 7-36.

[6] BOUCHITTE G.: "Représentation intégrale de fonctionnelles convexes sur un espace de mesures. Convergence". *Ann. Univ. Ferrara.*, 33, 1987, p 113-156.

[7] BOUCHITTE G., SUQUET P.: "Charges limites, Plasticité et homogénéisation: le cas d'un bord chargé". *C. R. Acad. Sc. Paris*, I, 305, 1987, p 441-444.

[8] BOUCHITTE G., VALADIER M: "Multi-fonctions s.c.i. et régularisée s.c.i. essentielle. Fonctions de mesure dans le cas sous linéaire". in *Proceedings Congrès Franco-Québecois Analyse non linéaire*. Gauthier Villars. Paris. 1989.

[9] BOUCHITTE G., VALADIER M.: "Integral representation of convex functionals on a space of measures". J. Functional Analysis, 80, 1988, p 398-420.

[10] DAL MASO G.: "Integral representation on BV(Ω) of Γ-limits of variational integrals". *Manuscripta Math.*, 30, 1980, p 387-416.

[11] DE BUHAN P.: *Approche fondamentale du calcul à la rupture des ouvrages en sols renforcés*. Thèse d'Etat. Paris. 1986.

[12] DE GIORGI E.: "Convergence problems for functionals and operators" in *Recent Methods in Nonlinear Analysis*. Pitagora. Bologna. 1978.p 131-188.

[13] DEMENGEL F., TANG QI : "Homogénéisation en Plasticité". *C. R. Acad. Sc. Paris*, I, 303, 1986, p 339-342.

[14] MARCELLINI P.: "Periodic solutions and homogenization of nonlinear variational problems". *Annali Matematica Pura Appl.*, 117, 1978, p 139-152.

[15] SUQUET P.: "Un espace fonctionnel pour les équations de la Plasticité". *Ann. Fac. Sc. Toulouse*. 1, 1979, p 77-87.

[16] SUQUET P.: "Analyse limite et homogénéisation". *C. R. Acad. Sc. Paris*, II, 296, 1983, p 1355-1358.

[17] TEMAM R.: *Mathematical problems in Plasticity*. Gauthier Villars. Paris. 1985.

[18] TURGEMAN S., PASTOR J.: "Comparaison des charges limites d'une structure hétérogène et homogénéisée". *J. Méca. Th. Appl.*, 6, 1987, p 121-143.

APPENDIX
Layered materials

A.1 Strength domain of a layered material with two constituents

Consider a two dimensional layered medium with two constituents denoted 0 and 1, with volume fractions v_0 and v_1. The strength domain of each constituent is defined by :

(A.1) $$\sigma_{11}^2 + 2\sigma_{12}^2 + \sigma_{22}^2 \leq k^2(x) ,$$

where $k(x) = k_0$ or k_1, and $k_0 \leq k_1$.

The two layers being infinite in the x_1 direction, the microscopic stress fields can be assumed to depend only on x_2. Therefore it results from the equilibrium equations that σ_{12} and σ_{22} are constant and equal to their average:

(A.2) $$\sigma_{12} = \Sigma_{12} , \sigma_{22} = \Sigma_{22} .$$

The microscopic yield condition (A.1) now reads as:

$$|\sigma_{11}(x_2)| \leq \left(k^2(x) - 2\Sigma_{12}^2 + \Sigma_{22}^2 \right)^{1/2},$$

i.e.

$$|\Sigma_{11}| \leq v_0 \left(k_0^2 - 2\Sigma_{12}^2 + \Sigma_{22}^2 \right)^{1/2} + v_1 \left(k_1^2 - 2\Sigma_{12}^2 + \Sigma_{22}^2 \right)^{1/2}.$$

For biaxial stress states ($\Sigma_{12} = 0$, Σ_{11} and $\Sigma_{22} \neq 0$) the above macroscopic strength domain is delimited by fourth order curves, and by the two straight lines $\Sigma_{22} = \pm k_0$ (see figure A.1). There is a *"weakest link" effect* since the strength in the direction orthogonal to the layers is always equal to the smaller strength of the two materials, irrespective of the volume fraction of the constituents.

However in the direction of the layers the strengthening is effective: for instance the strength in uniaxial tension in the direction of the layers is the arithmetic mean of the constituents strengths $v_0 k_0 + v_1 k_1$.

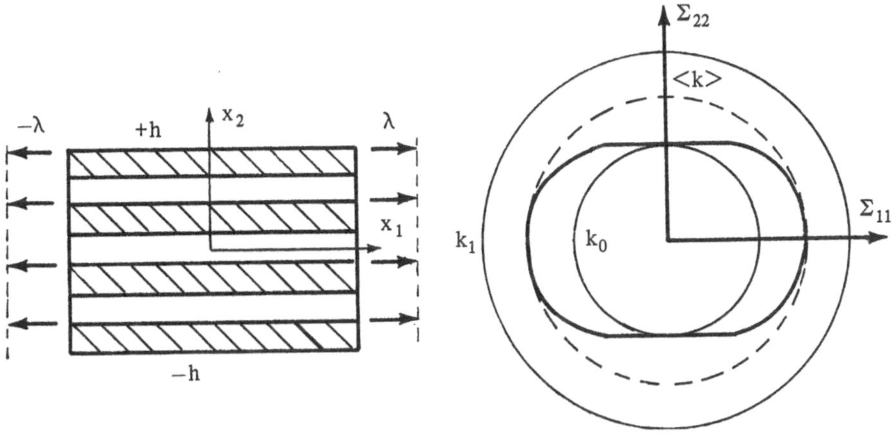

Figure A.1

A.2 An elementary illustration of why the A.V.P. fails

Now consider a rectangular block of this layered material, submitted to uniform tractions parallel to the layers direction:

$$\text{(A.3)} \qquad \begin{cases} \text{for } x_1 = \pm L, \; \sigma_{11} = \pm \lambda, \; \sigma_{12} = 0, \\ \text{for } x_2 = \pm h, \; \sigma_{22} = 0, \; \sigma_{12} = 0. \end{cases}$$

It is readily seen that :

$$\lambda^\epsilon = k_0 = \mathrm{Inf}\{ k_0 , k_1 \} \quad ,$$

since $| \sigma_{11}(x) | \le k_0$ a.e.x in $\Omega_0^\epsilon = \left\{ x \in \Omega \ , \ P^\epsilon(x) = P_0 \right\}$, $\sigma_{11} = \pm \lambda$ on $x_1 = \pm L$, and since $\partial\Omega_0^\epsilon$ intersects $\{ x_1 = \pm L\}$.

It is also readily checked that:

$$\mathrm{Sup}\{ \lambda | \ \Sigma \in P^{hom},\ \mathrm{div}(\Sigma) = 0,\ \Sigma \text{ satisfies (A.3)}\} = v_0 k_0 + v_1 k_1 .$$

Therefore $\lim(\lambda^\epsilon)$ is different from the result provided by the A.V.P..

A.3 Application of Theorem 1

We apply the general result provided by (3.42) for the layered material pictured on Figure A.1 (with strong layers of material 1 at top and bottom of the body), the sets A and B read as:

$$A = \partial\Omega \ , \ B = \{ x_1 = \pm L\} \ .$$

Therefore Remark 2 imply that :

$$C(x) = P_0 n(x) \text{ for every } x \text{ in } B.$$

We can state a more specific result, valid for any geometry of the layered material:

$$C(x) = P_0 n(x) \text{ for every } x \text{ in } \Gamma_1 .$$

To prove this affirmation we note that, on A - B, material 1 is exclusively present, and therefore that $n(x) = e_2$, where e_2 gives the direction orthogonal to the layering. According to (A.2) we have:

$$P^{hom} e_2 = \{ (\Sigma_{12}, \Sigma_{22}) \in \mathbb{R}^2,\ \exists\ \sigma(y_2) \in P(y) \text{ a.e. } y \text{ in } Y \ , \ \langle\sigma\rangle = \langle\Sigma\rangle,\ \sigma_{12} = \Sigma_{12},\ \sigma_{22} = \Sigma_{22} \ \}.$$

Note that:

$$P^0 e_2 = \left\{ (\sigma_{12}, \sigma_{22}) \in \mathbb{R}^2,\ \exists\ \sigma_{11} \text{ such that } (\sigma_{11}, \sigma_{12}, \sigma_{22}) \in P^0 \right\}.$$

Therefore, under the assumption $P^0 \subset P^1$, we have $P^{hom} e_2 = P^0 e_2$. We conclude by means of proposition 2, since:

$$(P^0 \cap P^1) e_2 = P^0 e_2 \ , \text{ and } C(x) \cap C^{hom}(x) = P^0 e_2 .$$

We have established the following result: *for a layered material with two constituents such that $P^0 \subset P^1$, the strength on the loaded boundary is ruled by the strength of the weakest constituent.*

We immediately conclude that, in the specific example under consideration, λ^{hom} given by (3.15), or equivalently by the primal characterization (1.10), is equal to k_0, i.e. to the limit of λ^ϵ. Q.E.D.

Guy BOUCHITTE
U.T.V.
Avenue de l'Université
BP 132. 83957. LA GARDE. Cedex
FRANCE

Pierre SUQUET
L.M.A.
31 Chemin Joseph Aiguier
13402. MARSEILLE. Cedex 09.
FRANCE

Some remarks on Γ-convergence and least squares method

ENNIO DE GIORGI

In the study of semicontinuity, relaxation, and Γ-convergence problems, few attention has been devoted, up to now, to questions concerning functionals arising in the study of differential equations or systems by the method of least squares. I think that a systematic study of these functionals could lead to interesting results, as, for instance, a reasonable "variational" definition of "weak solutions" of differential equations or systems.

Let us begin with a very simple example. For every $\Omega \subset\subset \mathbf{R}^{n+1}$, for every Lipschitz continuous function u defined on Ω, and for every $g \in L^2_{loc}(\mathbf{R}^n)$, $a_i \in L^2_{loc}(\mathbf{R}^{n+1})$, let us consider the functional

$$F(u,\Omega) = \int_\Omega \left(\frac{\partial u}{\partial t} - \sum_{i=1}^n a_i \frac{\partial u}{\partial x_i} \right)^2 d\mathcal{H}^{n+1}(x,t) \quad + \\ + \int_{\Omega_0} (u(x,0) - g(x))^2 d\mathcal{H}^n(x),$$

where $\Omega_0 = \{x \in \mathbf{R}^n : (x,0) \in \Omega\}$ and $\mathcal{H}^\alpha$ is the α-dimensional Hausdorff measure. For every $u \in L^2(\Omega) \setminus Lip(\Omega)$ we set $F(u,\Omega) = +\infty$.

It is clear that any function u satisfying the condition $F(u,\Omega) = 0$ for every $\Omega \subset\subset \mathbf{R}^{n+1}$ can be considered as a strong solution of the equation

$$(1) \qquad \frac{\partial u}{\partial t} = \sum_{i=1}^n a_i \frac{\partial u}{\partial x_i},$$

satisfying the condition

$$(2) \qquad u(x,0) = g(x).$$

Let us denote the relaxed functional of F by $\overline{F} = sc^-(L^2(\Omega))F$. It is reasonable to consider all functions $u \in L^2_{loc}(\mathbf{R}^{n+1})$, such that $\overline{F}(u,\Omega) = 0$ for every $\Omega \subset\subset \mathbf{R}^{n+1}$, as "variational weak solutions" of problem (1), (2).

Remark 1. It would be interesting to compare this notion of weak solution (possibly localized by considering only open sets $\Omega \subset\subset A$, where A is a suitable open subset of $\mathbf{R}^{n+1}$) with other notions of weak solutions existing in the literature (see, for instance, [2], [3], [5], [7], and the bibliography therein).

Remark 2. More generally, we can consider functionals of the form

$$F(u,\Omega) = \int_\Omega (L(u))^2 d\mathcal{H}^{n+1} + \int_{\Omega\cap S} (E(u))^2 d\mathcal{H}^n,$$

where L and E are differential operators and S is a suitable surface in $\mathbf{R}^{n+1}$. In this way it is not difficult to define, within a unified framework, the notion of "weak solution" of Cauchy, Dirichlet, Neumann, and other problems.

A simple and interesting example was studied by S. Mortola, who considered the functional

$$F_p(u,\Omega) = \int_\Omega \left(\frac{\partial u}{\partial t} - \frac{\sin px_1}{|\sin px_1|}\frac{\partial u}{\partial x_2}\right)^2 d\mathcal{H}^3(x,t) \quad + \\ + \int_{\Omega_0} (u(x,0) - g(x))^2 d\mathcal{H}^2(x),$$

where p is an arbitrary real number different from 0 and Ω is an open subset of $\mathbf{R}^3$.

It is easy to see that, for every $g \in L^2_{loc}(\mathbf{R}^2)$, the "variational weak solution" of equation (1) with condition (2) is given by the formula

$$u_p(x_1,x_2,t) = g\left(x_1, x_2 + t\frac{\sin px_1}{|\sin px_1|}\right).$$

If we take the limit of the functions u_p as $p \to +\infty$, we see that this limit exists only in the weak topology of $L^2_{loc}(\mathbf{R}^3)$ and it is given by the formula

$$u_\infty(x_1,x_2,t) = \frac{g(x_1,x_2+t) + g(x_1,x_2-t)}{2},$$

whereas there is no convergence in the strong topology of $L^2_{loc}(\mathbf{R}^3)$.

Let us consider now the functional $F_\infty = \Gamma(L^2_{loc}(\Omega)^-)\lim_{p\to+\infty} F_p$. We may expect that, for a Lipschitz continuous u and for $\Omega_0 = \emptyset$, we have

$$F_\infty(u,\Omega) = \int_\Omega \left(\left|\frac{\partial u}{\partial t}\right|^2 + \left|\frac{\partial u}{\partial x_2}\right|^2 \right) d\mathcal{H}^3(x,t),$$

and, in general, this functional does not vanish for $u = u_\infty$, even if u_∞ happens to be Lipschitz continuous. Therefore, it seems reasonable to study the functional $w{-}F_\infty = \Gamma(w{-}L^2_{loc}(\Omega)^-)\lim_{p\to+\infty} F_p$, where $w{-}L^2_{loc}(\Omega)$ denotes the weak topology of $L^2_{loc}(\Omega)$. This functional vanishes for $u = u_\infty$, but its behaviour with respect to Ω exhibits some interesting features which should be better understood. For instance, the following conjectures should be analysed.

Conjecture 1. For some functions u we have that $w{-}F_\infty(u,\Omega)$ is not subadditive with respect to Ω.

Conjecture 2. For $\Omega_0 = \emptyset$ and u Lipschitz continuous we have

$$w{-}F_\infty(u,\Omega) = $$
$$= \inf_{v\in Lip(\Omega)} \left\{ \int_\Omega \left[\left(\frac{\partial u}{\partial t} + \frac{\partial v}{\partial x_2}\right)^2 + \left(\frac{\partial u}{\partial x_2} + \frac{\partial v}{\partial t}\right)^2 \right] d\mathcal{H}^3(x,t) \right\}.$$

More generally one could consider periodic functions $a_i(x,t)$ and functionals of the form

$$F_p(u,\Omega) = \int_\Omega \left(\frac{\partial u}{\partial t} - \sum_{i=1}^n a_i(px,pt)\frac{\partial u}{\partial x_i} \right)^2 d\mathcal{H}^{n+1}(x,t) \quad +$$
$$+ \int_{\Omega_0} (u(x,0) - g(x))^2 d\mathcal{H}^n(x),$$

and discuss the following conjecture.

Conjecture 3. The following conditions are equivalent:

(i) both the Γ-limits

$$F_\infty = \Gamma(L^2_{loc}(\Omega)^-)\lim_{p\to+\infty} F_p, \qquad w{-}F_\infty = \Gamma(w{-}L^2_{loc}(\Omega)^-)\lim_{p\to+\infty} F_p$$

exist and $F_\infty = w{-}F_\infty$;

(ii) for every $g \in L^2_{loc}(\mathbf{R}^n)$ there exists a unique function u_∞ satisfying the condition

$$F_\infty(u_\infty, \Omega) = 0 \quad \text{for every } \Omega \subset\subset \mathbf{R}^{n+1};$$

(iii) for every $u \in L^2_{loc}(\mathbf{R}^{n+1})$ we have that $w{-}F_\infty(u, \Omega)$ is subadditive with respect to Ω.

Remark 3. Conjecture 3 remains an interesting open problem even under the additional hypothesis that the functions a_i are smooth and do not depend on the variable t.

Remark 4. These conjectures should be compared with the results obtained by L. Tartar (see, [11], [12]) and by other authors (see [1], [8]).

Remark 5. The example given by Mortola shows that, in the study of Γ-limits of functionals arising from the study of differential equations by the "method of least squares", it may sometimes happen that Γ-limits of measures are not measures. This fact, pointed out by Mortola in a special case, has, in general, an important meaning that should be further investigated.

Remark 6. Besides the functional $F(u, \Omega)$, we can consider, for every $p \geq 1$, the functionals

$$\int_\Omega \left(\frac{\partial u}{\partial t} - \sum_{i=1}^{n} a_i \frac{\partial u}{\partial x_i} \right)^p d\mathcal{H}^{n+1}(x,t) + \int_{\Omega_0} (u(x,0) - g(x))^p d\mathcal{H}^n(x)$$

and their relaxed functionals and Γ-limits in the L^p-spaces. In the case $p = 1$ we can also consider the relaxed functionals and Γ-limits in the space of measures endowed with the weak topology.

Another interesting case, where the Γ-limits of measures are probably not measures, can be obtained by considering the approximation of the perimeter functional given by a theorem by Modica and Mortola (see [10]). A particular case of this theorem is the following result:

If Ω is an open subset of $\mathbf{R}^n$, p a positive real number, and

$$F_p(u, \Omega) = \begin{cases} \displaystyle\int_\Omega \left(\frac{|\nabla u|^2}{p} + p(1 - \cos u) \right) dx & \forall u \in W^{1,1}(\Omega), \\ +\infty & \forall u \in L^1(\Omega) \setminus W^{1,1}(\Omega) \end{cases}$$

then

$$\Gamma(L^1(\Omega)^-)\lim_{p\to+\infty} F_p(u,\Omega) < +\infty$$

$$\text{if and only if} \quad u \in BV(\Omega) \quad \text{and} \quad \cos u = 1 \quad \text{a.e. in} \quad \Omega;$$

in this case

$$\Gamma(L^1(\Omega)^-)\lim_{p\to+\infty} F_p(u,\Omega) = 8\sqrt{2}\int_\Omega |Du|$$

where $\int_\Omega |Du|$ denotes the total variation on Ω of the vector measure Du (see Miranda [9]).

For every $\lambda > 0$, let G_p be the functional defined by

$$G_p(u,\Omega) = \int_\Omega \left[\left(\frac{\Delta u}{p} - p\sin u\right)^2 + \lambda\right]\left[\frac{|\nabla u|^2}{p} + p(1-\cos u)\right]dx,$$

if $u \in W^{2,1}(\Omega)$, and by $G_p(u,\Omega) = +\infty$, if $u \in L^1(\Omega)\setminus W^{2,1}(\Omega)$. It would be interesting to verify the following conjectures, for which we need the definitions of regular manifold and regular manifold with regular boundary in an open subset Ω of $\mathbf{R}^n$.

Definition 1. Let Ω be an open subset of $\mathbf{R}^n$ and let $h \in \mathbf{N}$, $h \le n$. For $\alpha \in \mathbf{N}$, $\alpha = \infty$, or $\alpha = \omega$, we define $V_hC^\alpha(\Omega)$ as the class of all sets $E \subset \mathbf{R}^n$ such that $E \cap \Omega = \overline{E} \cap \Omega$ and for every $x \in E \cap \Omega$ there exist two open sets $A \subset \mathbf{R}^n$, $B \subset \mathbf{R}^h$ and two functions $\varphi \in (C^\alpha(B))^n$, $\psi \in (C^\alpha(A))^h$ such that

$$x \in A, \quad \psi(\varphi(y)) = y \quad \text{for every } y \in B, \quad E \cap A = \varphi(B).$$

Moreover, we define $V_hBC^\alpha(\Omega)$ as the class of all sets $E \subset \mathbf{R}^n$ such that $E \cap \Omega = \overline{E} \cap \Omega$ and for every $x \in E \cap \Omega$ there exist two open sets $A \subset \mathbf{R}^n$, $B \subset \mathbf{R}^h$, two functions $\varphi \in (C^\alpha(B))^n$, $\psi \in (C^\alpha(A))^h$ and $z \in \mathbf{R}^h$, such that

$$x \in A, \quad \psi(\varphi(y)) = y \quad \text{for every } y \in B,$$
$$E \cap A = \{\varphi(y) : y \in B,\ \langle y, z\rangle \ge 0\}.$$

In other words, $E \in V_hC^\alpha(\Omega)$ (resp. $E \in V_hBC^\alpha(\Omega)$) is simply a set, closed in Ω, which is locally in one-to-one C^α-correspondence with a portion of the space $\mathbf{R}^h$ (resp. of the space $\mathbf{R}^h$ or of the half-space $\{y \in \mathbf{R}^h : \langle y, z\rangle \ge 0\}$).

Conjecture 4. If $E \in V_n BC^2(\mathbf{R}^n)$ and $u = 2\pi\chi_E$, there exists $k \in \mathbf{R}$, depending only on n, such that

$$\Gamma(L^1(\Omega)^-) \lim_{p\to+\infty} G_p(u,\Omega) = c\lambda\mathcal{H}^{n-1}(\partial E \cap \Omega) \quad + k \int_{\partial E\cap\Omega} H^2 d\mathcal{H}^{n-1},$$

where $c = 8\sqrt{2}$ and $H(x)$ is the mean curvature of ∂E at the point x.

Conjecture 5. There exist functions u such that, for every $\Omega \subset \mathbf{R}^n$, $\Gamma(L^1(\Omega)^-)\lim_{p\to+\infty} G_p(u,\Omega) < +\infty$, but for which this Γ-limit is not subadditive with respect to Ω. The typical case is when $u = 2\pi\chi_E$ and χ_E is the characteristic function of a set E whose boundary has some cusps.

Remark 7. The coefficient $\lambda > 0$ has been introduced in order to guarantee some coerciveness of the functionals G_p, that must be more coercive than the functionals F_p.

Remark 8. If f is an analytic function in $\mathbf{R}^n$, then for almost all values of $k \in \mathbf{R}$ the sets $E_k = \{x \in \mathbf{R}^n : f(x) > k\}$ satisfy the condition of conjecture 4.

Let us observe, finally, that it would be interesting to find a standard method to construct the approximating sequences for the Γ-limit of the functionals G_p. To this aim it is convenient to consider the function $MCM^*(2\pi\chi_E, p)$, defined as follows.

Definition 2. Let $p \in \mathbf{R}$, $f : \mathbf{R}^n \to \mathbf{R}$, $w : \mathbf{R}^n \times [0,+\infty[\to \mathbf{R}$. We set $w = MCM^*(f,p)$ if and only if, by definition:

$$\text{(3)} \qquad f \in L^1_{\text{loc}}(\mathbf{R}^n) \quad \text{and} \quad \int_{\mathbf{R}^n} |f(x)|e^{-\lambda|x|^2}dx < +\infty \text{ for every } \lambda > 0,$$

$$\text{(4)} \qquad \begin{cases} w \in C^\infty(\mathbf{R}^n\times]0,+\infty[) \text{ and} \\ \displaystyle\int_0^a dt \int_{\mathbf{R}^n} |w(x,t)|e^{-\lambda|x|^2}dx < +\infty \quad \forall\, \lambda > 0, 0 < a < +\infty, \end{cases}$$

$$\text{(5)} \qquad \lim_{t\to 0}\int_{\mathbf{R}^n} |w(x,t) - w(x,0)|e^{-\lambda|x|^2}dx = 0 \text{ for every } \lambda > 0,$$

$$\text{(6)} \qquad \frac{\partial w}{\partial t} = \Delta_x w - p^2 \sin w \quad \forall (x,t) \in \mathbf{R}^n\times]0,+\infty[,$$

(7) $$w(x,0) = f(x) \quad \text{for every } x \in \mathbf{R}^n.$$

Remark 9. It is well known that conditions (3), (4), and (5) ensure existence and uniqueness of the solution of Cauchy problem (6), (7) (see, for instance, [6]).

Conjecture 6. Let $t > 0$ be an arbitrary positive number and put $u_p(x,t) = MCM^*(2\pi\chi_E, p)(x, \frac{t}{p})$ for every $p > 0$. Under the hypotheses of conjecture 4, we have:

$$\Gamma(L^1(\Omega)^-) \lim_{p\to+\infty} G_p(2\pi\chi_E) =$$

$$= \lim_{p\to+\infty} \int_\Omega \left[\left(\frac{\Delta_x u_p}{p} - p\sin u_p \right)^2 + \lambda \right] \left[\frac{|\nabla_x u_p|^2}{p} + p(1-\cos u_p) \right] d\mathcal{H}^n(x),$$

and, in addition,

$$\lim_{p\to+\infty} u_p(x,t) = 2\pi\chi_E(x) \text{ for every } t > 0, \text{ and for a.e. } x \in \mathbf{R}^n.$$

Remark 10. As usual in the theory of semicontinuity and Γ-convergence, conjectures 1-6 could be weakened by requiring that they hold only for a rich family of open sets Ω, instead of for all relatively compact open sets Ω. This avoids trivial counterexamples.

REFERENCES

[1] Y. Amirat, K. Hamdache and A. Ziani: *Homogénéisation d'equations hyperboliques du premier ordre - Application aux milieux poreux*, preprint.

[2] M. Carriero, A. Leaci and E. Pascali: *Convergenza per l'equazione degli integrali primi associata al problema del rimbalzo elastico unidimensionale*, Ann. Mat. Pura Appl.(4) **33** (1983), 227-256.

[3] M. Carriero, A. Leaci and E. Pascali: *Sulle soluzioni delle equazioni alle derivate parziali del primo ordine in insiemi di perimetro finito con termine noto misura*, Rend. Sem. Mat. Univ. Padova **73** (1985), 63-87.

[4] G. Dal Maso and L. Modica: *A general theory of variational functionals*, in Topics in Functional Analysis 1980-81, Quaderni della Scuola Normale Superiore di Pisa, Pisa, 1982.

[5] R.J. DiPerna and P.L. Lions: *Ordinary differential equations, transport theory and Sobolev spaces*, Invent. Math. **98** (1989), 511-547.

[6] G. Folland: *Introduction to partial differential equations*, Princeton University Press, Princeton, N.J., 1976.

[7] A. Leaci: *Sulle soluzioni generalizzate di equazioni alle derivate parziali del primo ordine*, Note Mat. **4** (1984), 113-148.

[8] L. Mascarenhas: *A linear homogenization problem with time dependent coefficient*, Trans. Amer. Math. Soc. **281** (1984), 179-195.

[9] M. Miranda: *Distribuzioni aventi derivate misure. Insiemi di perimetro finito*, Ann. Scuola Norm. Sup. Pisa **18** (1964), 27-56.

[10] L. Modica and S. Mortola: *Un esempio di Γ^--convergenza*, Boll. Un. Mat. Ital. **14-B** (1977), 285-299.

[11] L. Tartar: *Remarks on homogenization*, Homogenization and Effective Moduli of Materials and Media, 228-246, the IMA Vol. in Math. and its Appl., vol. 1, Springer, 1986.

[12] L. Tartar: *Nonlocal effects induced by homogenization*, Partial Differential Equations and the Calculus of Variations, Essays in Honor of E. De Giorgi, edited by F. Colombini, A. Marino, L. Modica, S. Spagnolo, Birkhäuser, 1989.

Ennio De Giorgi
Scuola Normale Superiore
Piazza dei Cavalieri, 7
56100 PISA (ITALY)

Homogenization of Miscible Displacement in Unsaturated Aggregated Soils

Ulrich Hornung

Abstract

A double porosity model is derived for the transport of a solute in an unsaturated soil. The formal homogenization is carried out and the resulting macro-model is dealt with numerically. Comparisons of break-through curves are given for cases with different degrees of aggregation.

Contents

1 Introduction

During the last ten years many applications of the method of homogenization to flow and transport through porous media have been published (a survey article on this topic is [11]). The starting point are the papers by Keller [15] and Tartar [18] in which the derivation of Darcy's law as a macroscopic equation from Stokes' equations as a micro-model was shown. Improvements and generalizations of these results were given more recently, e.g. in [1] and [16]. The microstructure of a porous medium, namely the local geometry of the solid matrix and the pores, leads in a natural way to other applications of these mathematical techniques. Miscible displacement and its effect on the viscosity of the fluid has been studied in [17]. Adsorption and chemical reactions in the context of heterogeneous catalysis were dealt with in [12] and [13]. Penetration of solutes into the porous matrix leads to the phenomenon of chromatography, see [19]. Effects of poro-elasticity are usually described by Biot's model; it has been derived in [6] and and attempt to give a generalization to two-phase flow has been made in [3].

Whereas the above mentioned papers start from the pore-scale as the micro-scale, there are other important aspects by which one is lead to consider micro-structures on an intermediate scale. What we mean is a scale between the pores and the scale of interest for simulation. For oil reservoirs the concept of fractured rocks has been well known for a number of years [4]. Here, one distinguishes between two regimes of flow and transport, namely *blocks* and *fractures* or *fissures*. The two regimes differ in that they have porosities - and thus conductivities - on different orders of magnitude. Therefore, one talks about *double porosity models.* This aspect has been studied in the context of homogenization in [2]; an overview can be found in [7].

Almost independently from the fields of oil reservoir simulation and chemical engineering, in soil physics one has introduced the notion of *aggregated media*; see, e.g., [5] [8] [9]. Here, one makes the assumption that the soil consists of aggregates and a domain with larger pores between them. If, as a first approximation, water is *immobile* in the aggregates and *mobile* in the domain with large pores, the dynamics of immiscible displacement in such a medium differs significantly from that in an ordinary porous medium, see [10]. The well-posedness of initial boundary value problems for diffusion processes in such media was studied in [14].

In this paper we generalize the ideas of the paper [10] to the case of solute transport influenced by unsaturated flow in an aggregated soil. We make the assumption that the soil has a double porosity micro-structure. We assume that on the pore scale, there are two regimes with conductivities and diffusivities that differ by one or two orders of magnitude. We assume that in both regimes we have water flow governed by Darcy's law and the transport of a solute (a tracer, a dissolved salt, or an organic and nonreactive substance) which undergoes molecular diffusion and convection. For simplicity of the model we ignore hydrodynamic dispersion.

The purpose of the paper is to show that the method of homogenization is applicable to this class of problems. Furthermore, we demonstrate that nonstandard macro-models are being derived, and that the solutions of the model equations show a behavior which differs from those of well known models for miscible displacement.

2 The Micro-Model

Here, we make the usual assumption used in homogenization of differential equations in periodic structures: Z is the standard cell in three dimensional Euclidian space with volume $|Z| = 1$. Y_0 is an open subset of Z with regular boundary Γ such that $Y_0 \cup \Gamma \subset \text{int}\,(Z)$; and $Y_1 = Z \setminus (Y_0 \cup \Gamma)$ is assumed to be connected. Z, together with Y_0, Y_1, and Γ, are repeated periodically in space; this periodic structure is scaled by the factor ε and then put into a fixed domain Ω. In this way, we obtain two regimes Ω_0^ε and Ω_1^ε which are separated by the interface Γ^ε.

We use the following notations.

t, x	= time and space variable
ε	= scale parameter
Ω_i^ε	= regime with porosity of type i $(i = 0, 1)$
Γ^ε	= interface between the two regimes
$\vec{\nu}$	= normal vector on Γ^ε, outwards with respect to Ω_1^ε
θ_i^ε	= relative water content in regime i
ψ_i^ε	= hydraulic pressure head in regime i
$\vec{q}_i^\varepsilon$	= Darcy's velocity in regime i
k_i^ε	= hydraulic conductivity of regime i
v_i^ε	= concentration of the solute in regime i
$\vec{r}_i^\varepsilon$	= flow rate of the solute in regime i
d_i^ε	= diffusivity of the solute in regime i
$\vec{e}$	= unit vector positive upwards

We assume that the water flow obeys Darcy's law. Using the conservation of mass, continuity of the pressure head and the flux across the interface we get the following micro-model for thew water flow.

$$\begin{cases} \partial_t \theta_1^\varepsilon = -\nabla \cdot \vec{q}_1^\varepsilon & , x \in \Omega_1^\varepsilon \\ \vec{q}_1^\varepsilon = -k_1^\varepsilon(\nabla \psi_1^\varepsilon + \vec{e}) & , x \in \Omega_1^\varepsilon \\ \psi_1^\varepsilon = \psi_0^\varepsilon & , x \in \Gamma^\varepsilon \\ \vec{\nu} \cdot \vec{q}_1^\varepsilon = \vec{\nu} \cdot \vec{q}_0^\varepsilon & , x \in \Gamma^\varepsilon \\ \partial_t \theta_0^\varepsilon = -\nabla \cdot \vec{q}_0^\varepsilon & , x \in \Omega_0^\varepsilon \\ \vec{q}_0^\varepsilon = -\varepsilon^2 k_0^\varepsilon(\nabla \psi_0^\varepsilon + \vec{e}) & , x \in \Omega_0^\varepsilon \end{cases} \tag{1}$$

We assume that the transport of the solute is governed by Fick's law and has a convective part due to the water flow. Further, we make the assumption that for the solute there are neither chemical reactions, nor adsorption or penetration into the porous matrix. Using conservation of mass and continuity of the concentration and the flow rate on the interface we get the following micro-model for the solute.

$$\begin{cases} \partial_t(\theta_1^\varepsilon v_1^\varepsilon) = -\nabla \cdot \vec{r}_1^\varepsilon & , x \in \Omega_1^\varepsilon \\ \vec{r}_1^\varepsilon = -d_1^\varepsilon(\nabla v_1^\varepsilon + \vec{q}_1^\varepsilon v_1^\varepsilon) & , x \in \Omega_1^\varepsilon \\ v_1^\varepsilon = v_0^\varepsilon & , x \in \Gamma^\varepsilon \\ \vec{\nu} \cdot \vec{r}_1^\varepsilon = \vec{\nu} \cdot \vec{r}_0^\varepsilon & , x \in \Gamma^\varepsilon \\ \partial_t(\theta_0^\varepsilon v_0^\varepsilon) = -\nabla \cdot \vec{r}_0^\varepsilon & , x \in \Omega_0^\varepsilon \\ \vec{r}_0^\varepsilon = -\varepsilon^2 d_0^\varepsilon \nabla v_0^\varepsilon + \vec{q}_0^\varepsilon v_0^\varepsilon & , x \in \Omega_0^\varepsilon \end{cases} \tag{2}$$

3 The Homogenization Procedure

3.1 The Water

We expand θ_i^ε, ψ_i^ε, and k_i^ε $(i = 0, 1)$ into formal power series in terms of ε (here we drop the argument t for simplicity of the notation).

$$\begin{aligned} \theta_i^\varepsilon(x) &= \theta_i^0(x,y) + \varepsilon\theta_i^1(x,y) + \varepsilon^2\theta_i^2(x,y) \ldots \\ \psi_i^\varepsilon(x) &= \psi_i^0(x,y) + \varepsilon\psi_i^1(x,y) + \varepsilon^2\psi_i^2(x,y) \ldots \\ k_i^\varepsilon(x) &= k_i^0(x,y) + \varepsilon k_i^1(x,y) + \varepsilon^2 k_i^2(x,y) \ldots \end{aligned}$$

where the coefficient functions θ_i^j, ψ_i^j, and k_i^j are periodic with respect to the fast variable $y = \frac{x}{\varepsilon}$. The differential equations for θ_1^ε and ψ_1^ε give

$$\begin{aligned}
&\varepsilon^0 \partial_t \theta_1^0 + \varepsilon^1 \ldots = \varepsilon^{-2} \nabla_y \cdot (k_1^0 \nabla_y \psi_1^0) \\
&\quad + \varepsilon^{-1}(\nabla_y \cdot (k_1^0 \nabla_y \psi_1^1 + k_1^1 \nabla_y \psi_1^0 + k_1^0(\nabla_x \psi_1^0 + \vec{e})) + \nabla_x \cdot (k_1^0 \nabla_y \psi_1^0)) \\
&\quad + \varepsilon^0(\nabla_y \cdot (k_1^0 \nabla_y \psi_1^2 + k_1^1 \nabla_y \psi_1^1 + k_1^2 \nabla_y \psi_1^0 + k_1^0 \nabla_x \psi_1^1 + k_1^1(\nabla_x \psi_1^0 + \vec{e})) \\
&\quad + \nabla_x \cdot (k_1^0 \nabla_y \psi_1^1 + k_1^1 \nabla_y \psi_1^0 + k_1^0(\nabla_x \psi_1^0 + \vec{e}))) \\
&\quad + \varepsilon^1(\ldots) + \ldots, \ y \in Y_1
\end{aligned} \tag{3}$$

The two conditions for ψ_i and $\vec{q}_i$ on Γ^ε give

$$\varepsilon^0 \psi_1^0 + \varepsilon^1 \ldots = \varepsilon^0 \psi_0^0 + \varepsilon^1 \ldots, \ y \in \Gamma$$

and

$$\begin{aligned}
&\varepsilon^{-1} \vec{\nu} \cdot (k_1^0 \nabla_y \psi_1^0) \\
&\quad + \varepsilon^0 \vec{\nu} \cdot (k_1^0 \nabla_y \psi_1^1 + k_1^1 \nabla_y \psi_1^0 + k_1^0(\nabla_x \psi_1^0 + \vec{e})) \\
&\quad + \varepsilon^1 \vec{\nu} \cdot (k_1^0 \nabla_y \psi_1^2 + k_1^1 \nabla_y \psi_1^1 + k_1^2 \nabla_y \psi_1^0 + k_1^0 \nabla_x \psi_1^1 + k_1^1(\nabla_x \psi_1^0 + \vec{e})) \\
&\quad + \varepsilon^2 \ldots = \varepsilon^1 \vec{\nu} \cdot (k_0^0 \nabla_y \psi_0) + \varepsilon^2 \ldots, \ y \in \Gamma
\end{aligned} \tag{4}$$

The differential equations for θ_0^ε and ψ_0^ε give

$$\varepsilon^0 \partial_t \theta_0^0 + \varepsilon^1 \ldots = \varepsilon^0 \nabla_y \cdot (k_0^0 \nabla_y \psi_0^0) + \varepsilon^1 \ldots, \ y \in Y_0 \tag{5}$$

Now we compare the coefficients of the different ε-powers in these equations. The ε^{-2}-term in 3 gives

$$\nabla_y \cdot (k_1^0 \nabla_y \psi_1^0) = 0, \ y \in Y_1$$

hence

$$\psi_1^0(x, y) = \psi_1^0(x)$$

independently of $y \in Y_1$. Therefore, the ε^{-1}-term in 3 reduces to

$$\nabla_y \cdot (k_1^0 \nabla_y \psi_1^1) = -\nabla_y \cdot (k_1^0(\nabla_x \psi_1^0 + \vec{e})), \ y \in Y_1$$

and the ε^0-term in 4 gives

$$\vec{\nu} \cdot (k_1^0 \nabla_y \psi_1^1) = -\vec{\nu} \cdot (k_1^0 \nabla_x(\psi_1^0 + \vec{e})), \ y \in \Gamma$$

We introduce the *cell function* σ_j which are Z-periodic function of the fast variable y and satisfy the equations

$$\begin{cases} \nabla_y \cdot (k_1^0 \nabla_y \sigma_j) = -\nabla_y \cdot (k_1^0 \vec{e}_j), & y \in Y_1 \\ \vec{\nu} \cdot (k_1^0 \sigma_j) = -\vec{\nu} \cdot \vec{e}_j, & y \in \Gamma \end{cases}$$

where $\vec{e}_j$ is the j-th unit vector in space. Using these cell functions, we find the follwing representation for the function ψ_1^1 in terms of ψ_1^0.

$$\psi_1^1(x, y) = \sum_{j=1}^n \sigma_j(y)(\partial_{x_j} \psi_1^0(x) + e_j) + \psi_1^1(x), \ y \in Y_1$$

where $\psi_1^1(x)$ is independent of y and e_j is the j-th component of $\vec{e}$. The ε^0-term in 3 yields

$$\begin{aligned}
&\partial_t \theta_1^0 = \\
&\quad \nabla_y \cdot (k_1^0 \nabla_y \psi_1^2 + k_1^1 \nabla_y \psi_1^1 + k_1^0 \nabla_x \psi_1^1 + k_1^1(\nabla_x \psi_1^0 + \vec{e})) \\
&\quad + \nabla_x \cdot (k_1^0 \nabla_y \psi_1^1 + k_1^0(\nabla_x \psi_1^0 + \vec{e})), \ y \in Y_1
\end{aligned}$$

We integrate this identity over Y_1 and obtain

$$\begin{aligned}
&|Y_1| \partial_t \theta_1^0 = \\
&\quad \int_{Y_1} \nabla_y \cdot (k_1^0 \nabla_y \psi_1^2 + k_1^1 \nabla_y \psi_1^1 + k_1^0 \nabla_x \psi_1^1 + k_1^1(\nabla_x \psi_1^0 + \vec{e}))\, dy \\
&\quad + \int_{Y_1} \nabla_x \cdot (k_1^0 \nabla_y \psi_1^1 + k_1^0(\nabla_x \psi_1^0 + \vec{e}))\, dy
\end{aligned} \tag{6}$$

We integrate the first term on the right hand side by parts and obtain

$$\begin{aligned}
&\int_{Y_1} \nabla_y \cdot (k_1^0 \nabla_y \psi_1^2 + k_1^1 \nabla_y \psi_1^1 + k_1^0 \nabla_x \psi_1^1 + k_1^1(\nabla_x \psi_1^0 + \vec{e}))\, dy \\
&\quad = \int_{\partial Y_1} \vec{\nu} \cdot (k_1^0 \nabla_y \psi_1^2 + k_1^1 \nabla_y \psi_1^1 + k_1^0 \nabla_x \psi_1^1 + k_1^1(\nabla_x \psi_1^0 + \vec{e}))\, d\Gamma(y)
\end{aligned}$$

We have $\partial Y = \partial Z + \Gamma$; the boundary integral over ∂Z vanishes due the periodicity. For the integral over Γ have have a look at the ε^1-term in 4 and get

$$\int_{\partial Y_1} \vec{\nu} \cdot (k_1^0 \ldots)\, dy = \int_\Gamma \vec{\nu} \cdot (k_0^0 \nabla_y \psi_0^0)\, d\Gamma(y)$$

For the first part of the second term in 6 we get

$$\begin{aligned}
&\int_{Y_1} \nabla_x \cdot (k_1^0 \nabla_y \psi_1^1)\, dy \\
&\quad = \sum_{i,j=1}^n \partial_{x_i} \int_{Y_1} k_1^0 \partial_{y_i} (\sigma_j(y)(\partial_{x_j} \psi_1^0(x) + e_j) + \psi_1^1(x))\, dy \\
&\quad = \sum_{i,j=1}^n \partial_{x_i} \int_{Y_1} k_1^0 \partial_{y_i} \sigma_j(y)\, dy\, (\partial_{x_j} \psi_1^0(x) + e_j)
\end{aligned}$$

Therefore, we obtain from equation 6

$$\begin{aligned}
&|Y_1| \partial_t \theta_1^0 - \int_\Gamma \vec{\nu} \cdot (k_0^0 \nabla_y \psi_0^0)\, d\Gamma(y) \\
&\quad = \sum_{i,j=1}^n \partial_{x_i} \int_{Y_1} k_1^0 \partial_{y_i} \sigma_j(y)\, dy\, (\partial_{x_j} \psi_1^0(x) + e_j) + \nabla_x \cdot (\int_{Y_1} k_1^0\, dy\, (\nabla_x \psi_1^0 + \vec{e}))
\end{aligned} \tag{7}$$

The ε^0-term in equation 5 yields

$$\partial_t \theta_0^0 = \nabla_y \cdot (k_0^0 \nabla_y \psi_0^0), \ y \in Y_0$$

The boundary condition is

$$\psi_1^0 = \psi_0^0, \ y \in \Gamma$$

3.2 The Solute

From the previous section we have learnt that the water fluxes have formal powers series of the form

$$\vec{q}_1^\varepsilon = \varepsilon^0 \vec{q}_1^0 + \varepsilon^1 \ldots$$

and

$$\vec{q}_0^\varepsilon = \varepsilon^1 \vec{q}_0^0 + \varepsilon^2 \ldots$$

Therefore, the differential equations for v_1^ε give

$$\begin{aligned}
&\varepsilon^0 \partial_t(\theta_1^0 v_1^0) + \varepsilon^1 \ldots = \varepsilon^{-2} \nabla_y \cdot (d_1^0 \nabla_y v_1^0) \\
&\quad + \varepsilon^{-1}(\nabla_y \cdot (d_1^0 \nabla_y v_1^1 + d_1^1 \nabla_y v_1^0 + d_1^0(\nabla_x v_1^0 - \vec{q}_1^0 v_1^0)) + \nabla_x \cdot (d_1^0 \nabla_y v_1^0)) \\
&\quad + \varepsilon^0(\nabla_y \cdot (d_1^0 \nabla_y v_1^2 + d_1^1 \nabla_y v_1^1 + d_1^2 \nabla_y v_1^0 \\
&\quad\ + d_1^0(\nabla_x v_1^1 - \vec{q}_1^1 v_1^0 - \vec{q}_1^0 v_1^1) + d_1^1(\nabla_x v_1^0 - \vec{q}_1^0 v_1^0)) \\
&\quad\ + \nabla_x \cdot (d_1^0 \nabla_y v_1^1 + d_1^1 \nabla_y v_1^0 + d_1^0(\nabla_x v_1^0 - \vec{q}_1^0 v_1^0))) \\
&\quad + \varepsilon^1(\ldots) + \ldots, \ y \in Y_1
\end{aligned} \tag{8}$$

The two conditions for v_i and $\vec{r}_i$ on Γ^ε give

$$\varepsilon^0 v_1^0 + \varepsilon^1 \ldots = \varepsilon^0 v_0^0 + \varepsilon^1 \ldots, \ y \in \Gamma$$

and

$$\begin{aligned}
&\varepsilon^{-1} \vec{\nu} \cdot (d_1^0 \nabla_y v_1^0) \\
&\quad + \varepsilon^0 \vec{\nu} \cdot (d_1^0 \nabla_y v_1^1 + d_1^1 \nabla_y v_1^0 + d_1^0(\nabla_x v_1^0 - \vec{q}_1^0 v_1^0)) \\
&\quad + \varepsilon^1 \vec{\nu} \cdot (d_1^0 \nabla_y v_1^2 + d_1^1 \nabla_y v_1^1 + d_1^2 \nabla_y v_1^0 \\
&\quad\ + d_1^0(\nabla_x v_1^1 - \vec{q}_1^1 v_1^0 - \vec{q}_1^0 v_1^1) + d_1^1(\nabla_x v_1^0 - \vec{q}_1^0 v_1^0)) \\
&\quad + \varepsilon^2 \ldots = \varepsilon^1 \vec{\nu} \cdot (d_0^0 \nabla_y v_0) + \varepsilon^2 \ldots, \ y \in \Gamma
\end{aligned} \tag{9}$$

The differential equations for v_0^ε give

$$\varepsilon^0 \partial_t(\theta_0^0 v_0^0) + \varepsilon^1 \ldots = \varepsilon^0 \nabla_y \cdot (d_0^0 \nabla_y v_0^0 - \vec{q}_0^0 v_0^0) + \varepsilon^1 \ldots, \ y \in Y_0 \tag{10}$$

Now we compare the coefficients of the different ε-powers in these equations. The ε^{-2}-term in 8 gives

$$\nabla_y \cdot (d_1^0 \nabla_y v_1^0) = 0, \ y \in Y_1$$

hence

$$v_1^0(x, y) = v_1^0(x)$$

independently of $y \in Y_1$. Therefore, the ε^{-1}-term in 8 reduces to

$$\nabla_y \cdot (d_1^0 \nabla_y v_1^1) = -\nabla_y \cdot (d_1^0(\nabla_x v_1^0 - \vec{q}_1^0 v_1^0)), \ y \in Y_1$$

and the ε^0-term in 9 gives

$$\vec{\nu} \cdot (d_1^0 \nabla_y v_1^1) = -\vec{\nu} \cdot (d_1^0(\nabla_x v_1^0 - \vec{q}_1^0 v_1^0)), \ y \in \Gamma$$

In the same way as for the water flow we can express the function v_1^1 in terms of v_1^0.

$$v_1^1(x, y) = \sum_{j=1}^{n} \sigma_j(y)(\partial_{x_j} v_1^0(x) - q_{1,j}^0) + v_1^1(x), \ y \in Y_1$$

where $v_1^1(x)$ is independent of y and $q_{1,j}^0$ is the j-th component of $\vec{q}_1^0$. The ε^0-term in 8 yields

$$\begin{aligned}&\partial_t(\theta_1^0 v_1^0) = \\ &\quad \nabla_y \cdot (d_1^0 \nabla_y v_1^2 + d_1^1 \nabla_y v_1^1 + d_1^0(\nabla_x v_1^1 - \vec{q}_1^0 v_1^0) + d_1^1(\nabla_x v_1^0 - \vec{q}_1^0 v_1^0)) \\ &\quad +\nabla_x \cdot (d_1^0 \nabla_y v_1^1 + d_1^0(\nabla_x v_1^0 - \vec{q}_1^0 v_1^0)),\ y \in Y_1\end{aligned}$$

We integrate this identity over Y_1 and obtain

$$\begin{aligned}&|Y_1|\partial_t(\theta_1^0 v_1^0) = \\ &\quad \int_{Y_1} \nabla_y \cdot (d_1^0 \nabla_y v_1^2 + d_1^1 \nabla_y v_1^1 + d_1^0(\nabla_x v_1^1 - \vec{q}_1^0 v_1^0) + d_1^1(\nabla_x v_1^0 - \vec{q}_1^0 v_1^0))\, dy \\ &\quad + \int_{Y_1} \nabla_x \cdot (d_1^0 \nabla_y v_1^1 + d_1^0(\nabla_x v_1^0 - \vec{q}_1^0 v_1^0))\, dy \end{aligned} \tag{11}$$

We integrate the first term on the right hand side by parts and obtain

$$\begin{aligned}&\int_{Y_1} \nabla_y \cdot (d_1^0 \nabla_y v_1^2 + d_1^1 \nabla_y v_1^1 + d_1^0(\nabla_x v_1^1 - \vec{q}_1^0 v_1^0) + d_1^1(\nabla_x v_1^0 - \vec{q}_1^0 v_1^0))\, dy \\ &\quad = \int_{\partial Y_1} \vec{\nu} \cdot (d_1^0 \nabla_y v_1^2 + d_1^1 \nabla_y v_1^1 + d_1^0 \nabla_x v_1^1 + d_1^1(\nabla_x v_1^0 - \vec{q}_1^0 v_1^0))\, d\Gamma(y)\end{aligned}$$

We have $\partial Y = \partial Z + \Gamma$; the boundary integral over ∂Z vanishes due the periodicity. For the integral over Γ have have a look at the ε^1-term in 9 and get

$$\int_{\partial Y_1} \vec{\nu} \cdot (d_1^0 \ldots)\, dy = \int_\Gamma \vec{\nu} \cdot (d_0^0 \nabla_y v_0^0)\, d\Gamma(y)$$

For the first term in the second term in 11 we get

$$\begin{aligned}&\int_{Y_1} \nabla_x \cdot (d_1^0 \nabla_y v_1^1)\, dy \\ &\quad = \sum_{i,j=1}^n \partial_{x_i} \int_{Y_1} d_1^0 \partial_{y_i}(\sigma_j(y)(\partial_{x_j} v_1^0(x) - q_{1,j}^0) + v_1^1(x))\, dy \\ &\quad = \sum_{i,j=1}^n \partial_{x_i} \int_{Y_1} d_1^0 \partial_{y_i} \sigma_j(y)\, dy\ (\partial_{x_j} v_1^0(x) - q_{1,j}^0)\end{aligned}$$

Therefore, we obtain from equation 11

$$\begin{aligned}&|Y_1|\partial_t(\theta_1^0 v_1^0) - \int_\Gamma \vec{\nu} \cdot (d_0^0 \nabla_y v_0^0)\, d\Gamma(y) \\ &\quad = \sum_{i,j=1}^n \partial_{x_i} \int_{Y_1} d_1^0 \partial_{y_i} \sigma_j(y)\, dy\ (\partial_{x_j} v_1^0(x) - q_{1,j}^0) \\ &\quad +\nabla_x \cdot (\int_{Y_1} d_1^0\, dy\ (\nabla_x v_1^0 - \vec{q}_1^0 v_1^0))\end{aligned} \tag{12}$$

The ε^0-term in equation 10 yields

$$\partial_t \theta_0^0 = \nabla_y \cdot (d_0^0 \nabla_y v_0^0 - \vec{q}_0^0 v_0^0),\ y \in Y_0$$

The boundary condition is

$$v_1^0 = v_0^0,\ y \in \Gamma$$

4 The Macro-Model

Now we drop the superscript 0 and define the conductivity tensors

$$K_{1,ij} = \int_{Y_1} k_1(\partial_{y_i}\sigma_j + \delta_{ij})\, dy \text{ and } K_{0,ij} = \delta_{ij} k_0$$

and the diffusivity tensors

$$D_{1,ij} = \int_{Y_1} d_1(\partial_{y_i}\sigma_j + \delta_{ij})\, dy \text{ and } D_{0,ij} = \delta_{ij} d_0$$

where δ_{ij} is the Kronecker delta. In this way, we obtain from the preceeding two sections the following macro-model.

The water flow is governed by the following system.

$$\left\{\begin{array}{ll} |Y_1|\partial_t\theta_1 + I = -\nabla_x \cdot \vec{q}_1 & ,x \in \Omega \\ \vec{q} = -K_1(\nabla_x\psi_1 + \vec{e}) & ,x \in \Omega \\ I = -\int_\Gamma \vec{\nu} \cdot (K_0\nabla_y\psi_0)\, d\Gamma(y) & ,x \in \Omega \\ \partial_t\theta_0 = -\nabla_y \cdot \vec{q}_0 & ,x \in \Omega,\ y \in Y \\ \vec{q}_0 = -K_0\nabla_y\psi_0 & ,x \in \Omega,\ y \in Y \\ \psi_1 = \psi_0 & ,x \in \Omega,\ y \in \Gamma \end{array}\right.$$

The solute transport is described by the following equations.

$$\left\{\begin{array}{ll} |Y_1|\partial_t(\theta_1 v_1) + J = \nabla_x \cdot (D_1\nabla_x v_1 - \vec{q}_1 v_1) & ,x \in \Omega \\ J = -\int_\Gamma \vec{\nu} \cdot (D_0\nabla_y v_0)\, d\Gamma(y) & ,x \in \Omega \\ \partial_t(\theta_0 v_0) = \nabla_y \cdot (D_0\nabla_y v_0 - \vec{q}_0 v_0) & ,x \in \Omega,\ y \in Y \\ v_1 = v_0 & ,x \in \Omega,\ y \in \Gamma \end{array}\right.$$

Using integration by parts and the differential equations with respect to the fast variable y one gets for the exchange terms

$$I = \int_{Y_0} \nabla_y \cdot (K_0\nabla_y\psi_0)\, dy = \int_{Y_0} \partial_t\theta_0\, dy$$

and

$$J = \int_{Y_0} \nabla_y \cdot (D_0\nabla_y v_0 - \vec{q}_0 v_0)\, dy = \int_{Y_0} \partial_t(\theta_0 v_0)\, dy.$$

5 Numerical Calculations

As a simple test example we have chosen vertical infiltration into an aggregated soil, i.e., a one-dimensional problem with respect to the slow variable which we call z here. For simplicity of the numerical method we have assumed that the aggregates are balls in 3-D space; thus, we can simplify the local problems and write the differential equations in terms of the radial variable r. For the water flow we have the following initial boundary value problem.

$$\begin{cases} |Y_1|\partial_t\theta_1 + aI = -\partial_z q_1 & ,t>0,\ -L<z<0 \\ q_1 = -K_1(\partial_z\psi_1 + 1) & ,t>0,\ -L<z<0 \\ I = \int_{Y_0}\partial_t\theta_0 dy \ / \ \int_{Y_0} dy & ,t>0,\ -L<z<0 \\ \partial_t\theta_0 = -\nabla_y\cdot\vec{q_0} & ,t>0,\ -L<z<0,\ |y|<R \\ \vec{q_0} = -K_0\nabla_y\psi_0 & ,t>0,\ -L<z<0,\ |y|<R \\ \psi_1 = \psi_0 & ,t>0,\ -L<z<0,\ |y|=R \\ -K_1(\partial_z\psi_1+1) = -g & ,t>0,\ z=0 \\ \psi_1 = 0 & ,t>0,\ z=-L \\ \psi_1 = \psi_{1,init} & ,t=0,\ -L<z<0 \\ \psi_0 = \psi_{0,init} & ,t=0,\ -L<z<0,\ |y|<R \end{cases}$$

The boundary condition at the soil surface is a prescribed flux given by the function g which is shown in figure 1a. The boundary condition at the bottom of the column is a fixed pressure head. The initial conditions chosen are the steady state situation compatible with the boundary conditions for $t = 0$.

For the solute we get the following initial boundary value problem.

$$\begin{cases} |Y_1|\partial_t(\theta_1 v_1) + aJ = \partial_z\cdot(D_1\partial_z v_1 - q_1 v_1) & ,t>0,\ -L<z<0 \\ J = \int_{Y_0}\partial_t(\theta_0 v_0)dy \ / \ \int_{Y_0} dy & ,t>0,\ -L<z<0 \\ \partial_t(\theta_0 v_0) = \nabla_y\cdot(D_0\nabla_y v_0 - \vec{q_0}v_0) & ,t>0,\ -L<z<0,\ |y|<R \\ v_1 = v_0 & ,t>0,\ -L<z<0,\ |y|=R \\ -D_1\partial_z v_1 + v_1 q_1 = -f & ,t>0,\ z=0 \\ \partial_z v_1 = 0 & ,t>0,\ z=-L \\ v_1 = v_{1,init} & ,t=0,\ -L<z<0 \\ v_0 = v_{0,init} & ,t=0,\ -L<z<0,\ |y|<R \end{cases}$$

The boundary condition at the surface of the soil is a given solute flux given by the function f which is shown in figure 1b. The boundary condition at the bottom of the soil column is an outflow condition. The initial conditions for the solute concentration are 0 everywhere.

The relations between ψ, θ and K are given in the figures 2a und 2b.

The numerical method uses finite elements both for the slow variable z and the fast variable r. The discretization with respect to time is dealt with using a fully implicit scheme. The nonlinear systems of equations for each time step are solved using a Newton method.

The results of the numerical calculations are shown in figures 3 and 4. There we use the symbol $A = |Y_0|$ in order to describe the *degree of aggregation*. Figure 3 shows the water flux at the bottom of the soil column as a function of time. And Figure 4 shows the solute flux, i.e., the so called *break-through-curve*. It is evident from the plots that the output from the column is the earlier the larger the degree of aggragation is. The same is true for the solute flux.

References

[1] G. ALLAIRE *Homogenization of the Stokes flow in a connected porous medium* Asympt. Anal. **2** (1989) 203-222

[2] ARBOGAST T., DOUGLAS J., HORNUNG U. *Derivation of the double porosity model of single phase flow via homogenization theory* SIAM J. Math. Anal. **21** (1990)

[3] AURIAULT J.-L., LEBAIGUE O., BONNET G. *Dynamics of two immiscible fluids flowing through deformable porous media* TiPM **4** (1989) 105-128

[4] G. I. BARENBLATT, I. P. ZHELTOV, I. N. KOCHINA *Basic concepts in the theory of seepage of homogeneous liquids in fissured rocks (strata)* J. Appl. Math. Mech. **24** (1960) 1286-1303

[5] BARKER J. A. *Block-geometry functions characterizing transport in densely fissured media* J. Hydrol. **77** (1985) 263-279

[6] BURRIDGE R., KELLER J. *Poroelasticity equations derived from microstructure* J. Acoust. Soc. Amer. **70** (1981) 1140-1146

[7] DOUGLAS J., ARBOGAST T. *Dual-porosity models for flow in naturally fractured reservoirs* Cushman J. (Ed.) "Dynamics of Fluids in Hierarchical Porous Media" Academic Press (1990) Chapter VII, 177-221

[8] GENUCHTEN M. Th., WIERENGA P. J. *Mass transfer studies in sorbing porous media I: Analytical solutions II: Experimental evaluation with tritium* (^3H_2O) Soil Sci. Soc. Amer. J. **40** (1976) 473-480 **41** (1977) 272-282

[9] GENUCHTEN M. Th., WIERENGA P. J., O'CONNOR G. A. *Mass transfer studies in sorbing porous media III: Experimental evaluation with 2,4,5-T* Soil Sci. Soc. Amer. J. **41** (1977) 278-285

[10] HORNUNG U. *Compartment and micro-structure models for miscible displacement in porous media* Rocky Mountain J. Math., to appear

[11] HORNUNG U. *Applications of the homogenization method to flow and transport through porous media* Xiao Shutie (Ed.) "Flow and Transport in Porous Media. Summer School, Beijing 1988" World Scientific, Singapore, to appear

[12] U. HORNUNG, W. JÄGER *A model for chemical reactions in porous media* J. Warnatz, W. Jäger (Eds.) "Complex Chemical Reaction Systems. Mathematical Modeling and Simulation" Chemical Physics **47** Springer, Berlin (1987) 318-334

[13] U. HORNUNG, W. JÄGER *Diffusion, convection, adsorption, and reaction of chemicals in porous media* J. Diff. Equat., to appear

[14] HORNUNG U., SHOWALTER R. *Diffusion models for fractured media* J. Math. Anal. Applics. **147** (1990) 69-80

[15] J. B. KELLER *Darcy's law for flow in porous media and the two-space method* R. L. Sternberg (Ed.) "Nonlinear Partial Differential Equations in Engineering and Applied Sciences" Dekker (1980) 429-443

[16] MIKELIĆ A. *A convergence theorem for homogenization of two-phase miscible flow through fractured reservoirs with uniform fracture distribution* Applic. Anal. **33** (1989) 203-214

[17] MIKELIĆ A. *Homogenization of nonstationary Navier-Stokes equations in a domain with a grained boundary* Ann. Mat. Pura Appl., to appear

[18] L. TARTAR *Incompressible fluid flow in a porous medium - convergence of the homogenization process* E. Sanchez-Palencia (Ed.) "Non-Homogeneous Media and Vibration Theory" Lecture Notes in Physics **127** Springer, Berlin (1980) 368-377

[19] C. VOGT *A homogenization theorem leading to a Volterra integro - differential equation for permeation chromatography* Preprint **155** SFB 123 (1982)

Ulrich Hornung, *SCHI*, P.O. Box 1222, D-Neubiberg, Germany

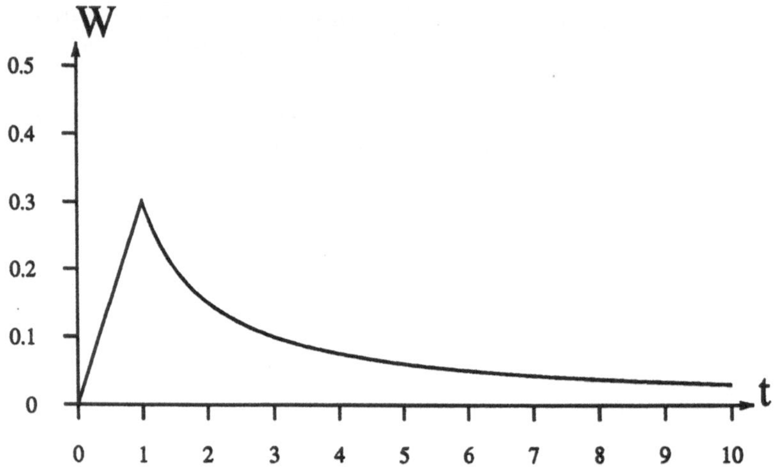

Figure 1a

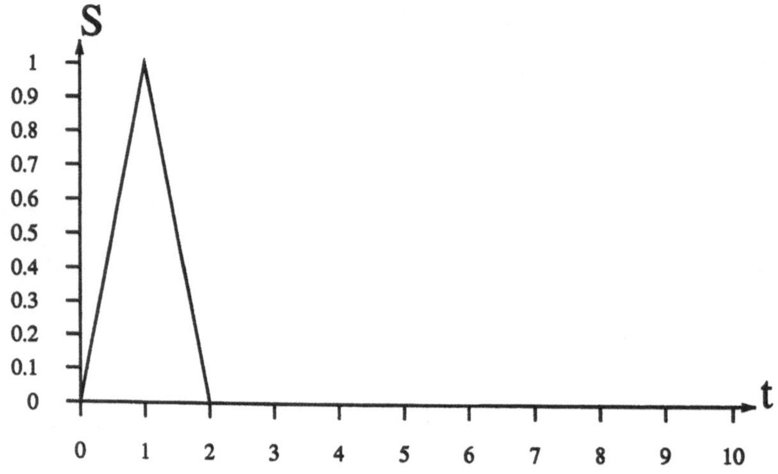

Figure 1b

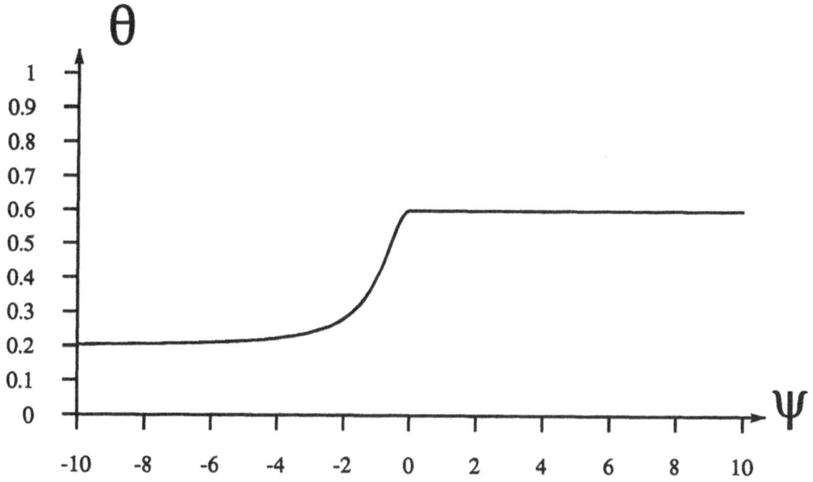

Figure 2a

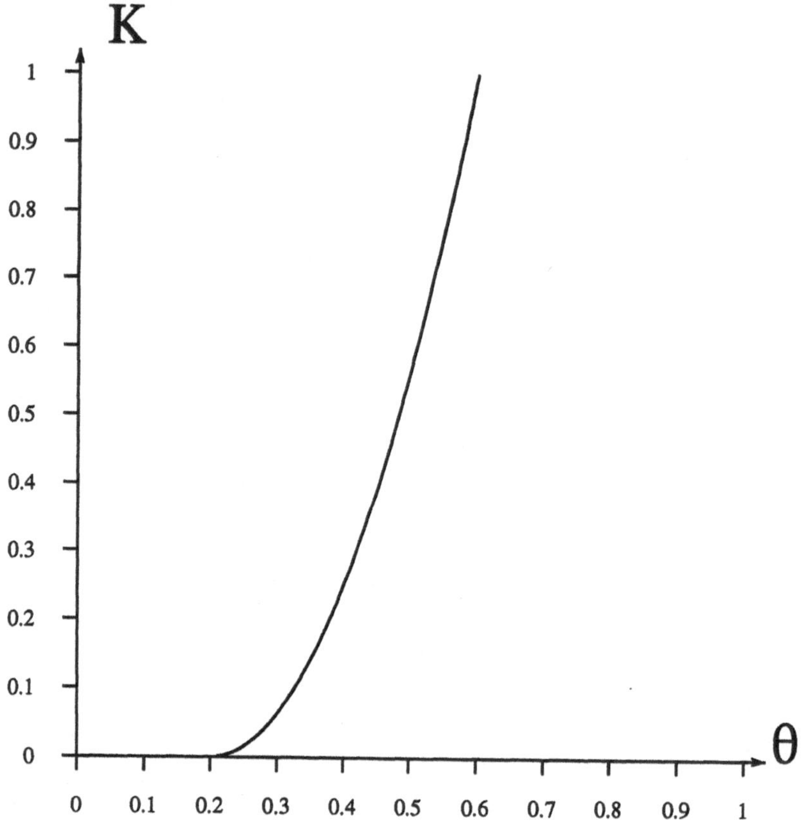

Figure 2b

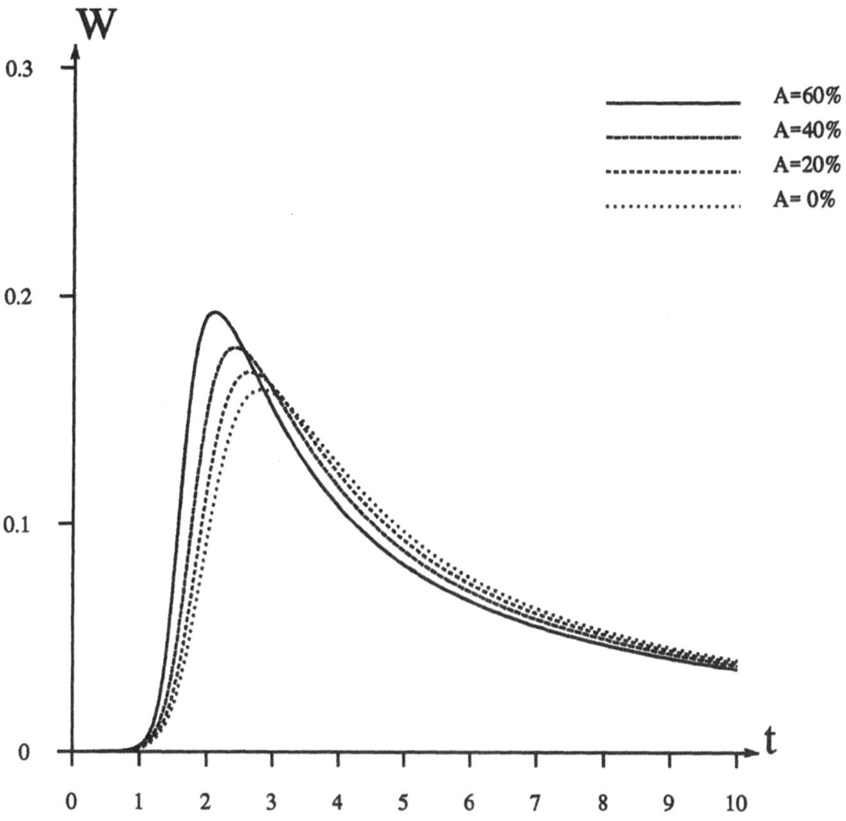

Figure 3

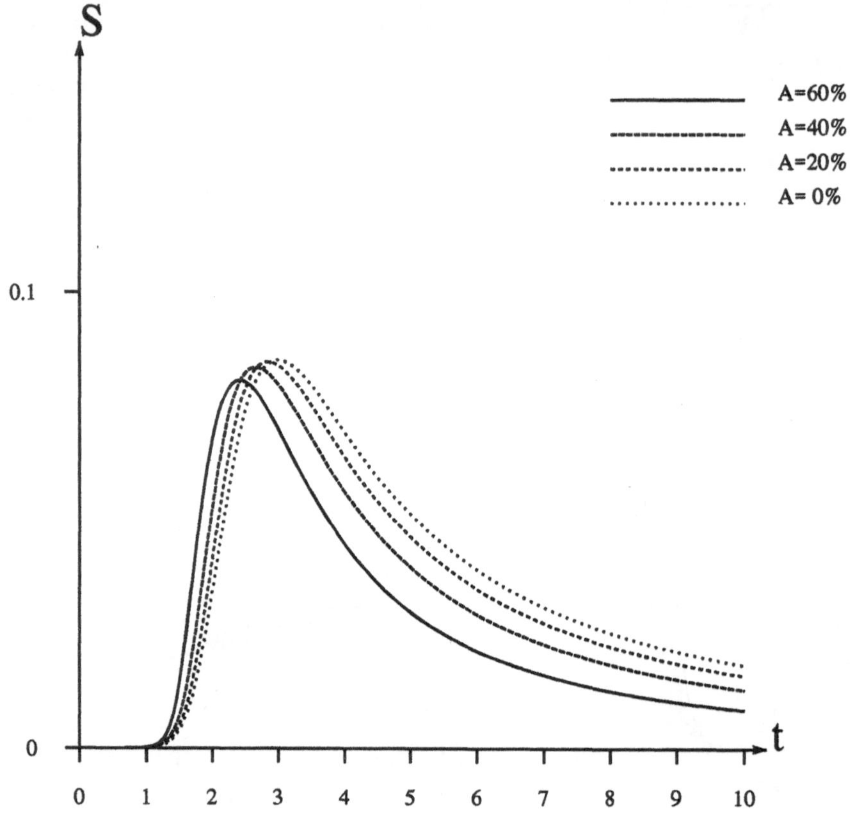

Figure 4

HOMOGENIZED MODELS OF COMPOSITE MEDIA

E.Ya. Khruslov

Abstract

In this paper we consider some homogenized models, which arise in consequence homogenization of boundary-value problems describing physical processes in highly inhomogeneous media. Such media take place, for example, in theory filtration, applied superconductivity,etc. Physical processes in them are described by both boundary-value problems in highly perforated domains and partial differential equations with rapidly oscillating coefficients, which do not satisfy conditions of uniform ellipticity or boundedness. The homogenization of such problems leads to unusual homogenized models (nonlocal,multiphase models, model with memory and others).

Boundary-value problems in highly perforated domains.

Let Ω be a bounded open set in space R^n ($n \geq 2$) and $F^{(s)}$ be a closed set in Ω, and $\Omega^{(s)}$ is the complement to $F^{(s)}$ in Ω : $\Omega^{(s)} = \Omega \setminus F^{(s)}$.We assume that set $F^{(s)}$ depends on the natural parameter s (s = 1,2,...) and for $s \to \infty$ the set $F^{(s)}$ becomes progressively more refined and more dense in the domain Ω.Exact conditions are following: for any ball B_ε^x in Ω of the radius ε,centered at the point $x \in \Omega$,if s is sufficiently large ($s \geq \hat{s}(\varepsilon)$), then intersection $F^{(s)} \cap B_\varepsilon^x$ is not empty and its complement $\Omega^{(s)} \cap B_\varepsilon^x$ also is not empty and connected.

Typical examples of such sets $F^{(s)}$ are so-colled 'fine-grained' sets consisting of a large number of small components :$F^{(s)} = \bigcup_{i=1}^{s} F_i^{(s)}$, but if dimension of space R^n ($F^{(s)} \subset R^n$) more then 2, then $F^{(s)}$ may be connected sets formed e.g. by intersecting wires.

We study the asymptotic behaviour for $s \to \infty$ solutions of boundary-value problems in domains $\Omega^{(s)} = \Omega \setminus F^{(s)}$. Our main aim is to state general conditions of convergence of these solutions to the functions determined in the domain Ω and to find the homogenized equations , describing these functions. The form of homogenized equations essentially depends on the geometric structure of the sets $F^{(s)}$ and the boundary condition on these sets. We shall demonstrate this for the case of the Dirichlet and Neumann boundary conditions on $F^{(s)}$ for the Helmholz equation in the domain $\Omega^{(s)} = \Omega \setminus F^{(s)}$. These problems lead to main qualitatively different types of homogenized models.

1. Dirichlet problem

Let us consider first Dirichlet boundary-value problem

$$(1.1) \qquad \Delta u^{(s)}(x) + \lambda u^{(s)}(x) = f(x) \qquad x \in \Omega.$$

$$(1.2) \qquad u^{(s)}(x) = 0. \qquad x \in \partial\Omega^{(s)}$$

where Δ is the Laplace operator in R^n, $f(x) \in L_2(\Omega)$, spectral parameter $\lambda \le 0$.

This problem has the unique solution $u^{(s)}(x)$ of Sobolev space $\overset{0}{H}{}^1(\Omega^{(s)})$. Continuing it by zero to the set $F^{(s)}$, we obtain the function $\hat{u}^{(s)}(x) \in \overset{0}{H}{}^1(\Omega)$. The sequence $\{\hat{u}^{(s)}(x), s = 1,2,\ldots\}$ is bounded in $\overset{0}{H}{}^1(\Omega)$ and therefore it is compact in $L_2(\Omega)$. To state the condition of its convergence and to describe its limit $u(x) \in L_2(\Omega)$ we introduce the functions $c_\varepsilon^{(s)}(x)$ on domain Ω:

$$c_\varepsilon^{(s)}(x) = \frac{\text{Cap }(\mathcal{F}^{(s)} \cap B_\varepsilon^x)}{\text{mes } B_\varepsilon^x}$$

where Cap denote Newton (for n = 2 Wiener) capacity, mes B_ε^x is volume of the ball B_ε^x, $\varepsilon > 0$.

The main result is following.

<u>THEOREM 1</u>

Let us assume ,that conditions

$$\text{(i)} \quad \lim_{\varepsilon\to 0} \underline{\lim}_{s\to\infty} c_\varepsilon^{(s)}(x) = \lim_{\varepsilon\to 0} \overline{\lim}_{s\to\infty} c_\varepsilon^{(s)}(x) = C(x)$$

at all Lebesgue point x of some measurable bounded function C(x) on Ω;

$$\text{(ii)} \qquad \overline{\lim}_{s\to\infty} c_\varepsilon^{(s)}(x) \le C < \infty , \; x \in \Omega, \; \varepsilon > 0$$

are valid.

Then the sequence $\{u^{(s)}(x)\}$ of solutions of the problem (1.1) - (1.2) (continued by zero to $F^{(s)}$) converges in $L_2(\Omega)$ to the solution u(x) of following homogenized problem

$$\text{(1.4)} \qquad \Delta u(x) - C(x)u(x) + \lambda u(x) = f(x), \qquad x \in \Omega$$

$$\text{(1.5)} \qquad u(x) = 0 \qquad x \in \partial\Omega$$

Conversely, if for any function $f(x) \in L_2(\Omega)$ the sequence $\{ \hat{u}^{(s)}(x)\}$ converges in $L_2(\Omega)$ to the solution of the problem (1.4) - (1.5), where the potential C(x) is measurable bounded function C(x) on Ω, then conditions (i)-(ii) are valid.

Thus, they are necessary and sufficient for strong convergence of the resolvent of the problem (1.1)-(1.2) to the resolvent of the problem (1.4) - (1.5).

The condition (i) permits to calculate potential C(x) explicitly in each concrete situation [1].

EXAMPLE 1

Let $F^{(s)}$ be connected set in $\Omega \subset R^3$ formed by thin threads of a diameter $d^{(s)} = A \exp\{-\frac{s^2}{a}\}$. Axes of these threads form in the space R^3 periodic net of the period $h^{(s)} = b/s$. Then

$$C(x) = \lim\left\{6\pi\, (h^{(s)})^{-2} \ln \frac{1}{d^{(s)}}\right\} = \frac{6\pi a}{b^2}$$

EXAMPLE 2

Let $F^{(s)}$ be random set consisting of s balls $F_i^{(s)}$ of the radius $r(s) = r / s$. These balls are distributed in the domain Ω with the probability density $\rho(x)$. In this case the solution $u^{(s)}(x)$ of the problem (1.1) - (1.2) is random and for $s \to \infty$ it converges in $L_2(\Omega)$ on probability to the nonrandom function u(x). This function is the solution of the homogenized problem (1.4)-(1.5), where potential C(x) is determined by formula

$$C(x) = 4\pi r \rho(x).$$

Note, that for Dirichlet boundary conditions the form of the homogenized equation (1.4) always remain the same, irrespective to the geometric structure of the sets $F^{(s)}$. The potential C(x),however,can be a distribution more exact in the measure [1]).A different situation arises in the case of the Neumann boundary condition.

2. Neumann problem

Let us consider in the domain $\Omega^{(s)} = \Omega \setminus F^{(s)}$ the following boundary-value problem

(2.1) $\Delta u^{(s)}(x) + \lambda u^{(s)}(x) = f(x)$, $x \in \Omega^{(s)}$

(2.2) $\frac{\partial u^{(s)}}{\partial \nu}(x) = 0$, $x \in \partial\Omega^{(s)}$

where $f(x) \in L_2(\Omega)$, $\frac{\partial}{\partial \nu}$ is normal derivative on boundary $\partial\Omega^{(s)}$ (for simplicity we assume, that for each s $\partial\Omega^{(s)}$ consists of smooth surfaces), spectral parameter λ is negative.

There exists the unique solution $u^{(s)}(x)$ of this problem and its norm in Sobolev space $H^1(\Omega^{(s)})$ is uniformly bounded with respect to s.

In the study of the asymptotic behaviour of $u^{(s)}(x)$ for $s \to \infty$ the first question to arise is following: for which domains $\Omega^{(s)}$ is it possible from the bounded sequence of of function $\{u^{(s)}(x) \in H^1(\Omega^{(s)})\}$ to select a subsequence, which converge in $L_2(\Omega^{(s)})$-norm to some function u(x), determined on Ω.The answer for this question may be given by mean of strong connectivity.

DEFINITION Let $\{B^{(s)}\}$ be a sequence of subdomains in Ω.We shall say, that this sequence satisfies the strong connectivity condition (or simply, $B^{(s)}$ are strongly connected) if for any sequence of functions { $v^{(s)}(x) \in C^1(B^{(s)})$, s = 1,2,..} and any M > 0 there exist subsets $B_M^{(s)} \subset B^{(s)}$ such, that for any x, $y \in B^{(s)} \setminus B_M^{(s)}$

$$|v^{(s)}| \leq M, \qquad |v^{(s)}(x) - v^{(s)}(y)| \leq M\,|x - y|$$

and for $s \geq \hat{s}(M)$

$$\text{mes}\,(B_M^{(s)}) = o\left(\frac{1}{M^2}\right) \cdot \|v^{(s)}\|_{H^1(B^{(s)})}$$

$$\|v^{(s)}\|^2_{L_2(B_M^{(s)})} = o(1) \cdot \|v^{(s)}\|^2_{H^1(B^{(s)})} \quad (M \to \infty)$$

More simpler sufficient condition of strong connectivity in term of the continuation of functions is given in [2].

EXAMPLES

1. Let $\Omega^{(s)} = \Omega \setminus F^{(s)}$ is complement of the union $F_i^{(s)} = \cup F_i^{(s)}$ of balls $F_i^{(s)}$ of small radius $r_i^{(s)}$ ($r_i^{(s)}$ 0 for s ∞).If this balls are distributed in domain Ω in such a way that the distance from each ball $F_i^{(s)}$ to rest balls $\underset{i \neq j}{\cup} F_j^{(s)}$ and boundary $\partial\Omega$ satisfies inequality

$$\text{dist}\,(F_i^{(s)}, \underset{i \neq j}{\cup} F_j^{(s)} \cup \Omega) \geq C\, r_i^{(s)}$$

where C does not depend on s, then domains $\{\Omega^{(s)}, s = 1,2,...\}$ are strongly connected.

2. Let $Q^{(s)}$ is union of round cylinders of radius $r^{(s)}$, which axes form in the space R^3 periodic cubic net of the period $h^{(s)} = b/s$. If $r^{(s)} \leq Ch^{(s)}$ $(C < 1/2)$, then domains $\{Q^{(s)}, s = 1,2,...\}$ and $\{R^3 \setminus Q^{(s)}, s = 1,2,...\}$ both satisfy the strong connectivity condition.

For strongly connected domains $\{\Omega^{(s)}, s = 1,2,...\}$ the above question finds an answer in the affirmative and therefore the homogenized equations for the problem (2.1) - (2.2) have the simplest form (local scalar models).On the other hand, for the domains $\Omega^{(s)}$, which do not satisfy the strong connectivity condition, the homogenized models prove more complicated (vector models, models with a memory, etc.).Let us consider the most typical cases.

2.1 Strongly connected domains

Introduce on each ball K_ε^x of the radius $\varepsilon > 0$ centered at the point $x \in \Omega$ the functional on vectors $l = (l_1, l_2, \ldots, l_n) \in R^n$:

$$A_{z\varepsilon}^{(s)}[l] = \inf_{v^{(s)}} \int_{K_\varepsilon^z \cap \Omega^{(s)}} \Big\{ |\nabla v^{(s)}(x)|^2 + \varepsilon^{-2-\theta} |v^{(s)}(x) - (x-z,l)|^2 \Big\} dx$$

where the infimum is taken among functions $v^{(s)}(x) \in$ $\in H^1(\Omega^{(s)} \cap K_\varepsilon^x)$, θ is arbitrary positive parameter ,$(\cdot,\cdot)$ denote the scalar product in R^n. The following representation is valid

$$A_{x\varepsilon}^{(s)}[l] = \sum_{i,k=1}^{n} a_{ik}(x,\varepsilon,s)\, l_i l_k$$

where $a_{ik}(x,\varepsilon,s)$ are components of a nonnegative tensor in the space R^n (conductivity tensor). This tensor and the Lebesgue measure $m(x,\varepsilon,s) = \text{mes}\,[K_\varepsilon^x \cap \Omega^{(s)}]$ are main local quantative characteristics of strongly connected domains $\Omega^{(s)}$ (in the neighbourhood of the point $x \in \Omega$)

The following theorem describes asymptotic behaviour of solution of Neumann boundary-value problem in such domains [2].

<u>THEOREM 2</u>

Let us assume, that conditions

$$\text{(i)} \qquad \lim_{\varepsilon\to 0} \underline{\lim}_{s\to\infty} \frac{a_{ik}(x,\varepsilon,s)}{\text{mes}\, K_\varepsilon^x} = \lim_{\varepsilon\to 0} \overline{\lim}_{s\to\infty} \frac{a_{ik}(x,\varepsilon,s)}{\text{mes}\, K_\varepsilon^x} = a_{ik}(x)$$

(ii) $$\lim_{\varepsilon\to 0}\lim_{s\to\infty}\frac{m(x,\varepsilon,s)}{\text{mes } K_\varepsilon^x} = m(x)$$

are fulfilled at all Lebesque point $x \in \Omega$ of measurable functions $a_{ik}(x)$ and $m(x)$ respectively, where tensor $\{ a_{ik}(x)\}_{i,k}^{n}$ is positive defined on Ω and $m(x) \geq m_0 \geq 0$.

Then the sequence of solution of the problem (2.1) - (2.2) converges in the norm $L_2(\Omega)$ to the solution of following homogenized problem

(2.3) $$\frac{1}{m(x)}\sum_{i,k=1}^{n}\frac{\partial}{\partial x_i}\left(a_{ik}(x)\frac{\partial u}{\partial x_k}\right) + \lambda\, u(x) = f(x)\ ,\ x \in \Omega$$

(2.4) $$\frac{\partial u}{\partial \nu_a}(x) = 0\ .\ x \in \partial\Omega$$

where $\frac{\partial}{\partial \nu_a} = \sum_{i,k=1}^{n} a_{ik}(x) \cos(\nu, x_i)\frac{\partial}{\partial x_i}$ is conormal derivative on $\partial\Omega$, corresponding to coefficients $a_{ik}(x)$. Converse theorem also valid. Thus in strongly connected domains conditions (i) - (ii) are necessary and sufficient for the convergence in $L_2(\Omega)$ - norm of solutions of the problem (2.1) - (2.2) to the solution of the problem (2.3) - (2.4).

REMARK

The tensor $\{ a_{ik}(x,\varepsilon,s)\}_{i,k}^{n}$, which enters in the condition (i), depends on a parameter $\theta > 0$ and limit conductivity tensor $\{ a_{ik}(x)\}_{i,k}^{n}$ does not depend on θ.

The condition (i) permits to calculate this tensor for many concrete situations. Particularly, if set $\mathcal{F}^{(s)}$ are systems of periodically (or locally periodically) arranged bodies, then condition (i) permits to reseive the well-known formula for functions $a_{ik}(x)$, using solutions of the cell problem.

2.2. Weakly connected domains.

Let us consider now domain $\Omega^{(s)} = \Omega \setminus F^{(s)}$ which do not satisfy the strong connectivity condition. Suppose that for any s, the domain $\Omega^{(s)}$ can be represented in such form:

$$\Omega^{(s)} = \bigcup_{r=1}^{m} \Omega_r^{(s)} \cup Q^{(s)}$$

where number $m \geq 2$ and does not depend on s; for any r the sequence of subdomains $\{ \Omega_r^{(s)}, s = 1,2,\dots\}$ satisfies the strong connectivity condition and mes $\{ \Omega_r^{(s)} \cap K_\varepsilon^x\} \geq C\varepsilon^n$; subdomains $\Omega_r^{(s)} (r = 1,\dots,m)$ do not intersect and are weakly connected between selves via the set $Q^{(s)}$, which Lebesque measure decrease to zero for $s \to \infty$.

The local quantitative characteristic of the weak connection between subdomains $\Omega_r^{(s)} (r = 1,2,\dots m)$ is introduced by mean of the functional on $t = \{t_1, t_2,\dots, t_m\} \in R^m$:

$$B_{x\varepsilon}^{(s)}[t] = \inf_{v^{(s)}} \int_{K_\varepsilon^x \Omega^{(s)}} \left\{ |\nabla v^{(s)}|^2 + \varepsilon^{-2-\theta} \sum_{r=1}^{m} |v^{(s)} - t_r|^2 \chi_r^{(s)} \right\} dx$$

where infimum is taken among functions $v^{(s)} \in H^1(\Omega^{(s)} \cap K_\varepsilon^x)$, $\chi_r^{(s)}$ are characteristic functions of subdomains $\Omega_r^{(s)}$.

The following representation is valid:

$$B_{x\varepsilon}^{(s)}[t] = \sum_{r,q=1}^{m} b_{rq}(x,\varepsilon,s) t_r t_q$$

where $b_{rq}(x,\varepsilon,s)$ are elements of so-called connection matrix. If for sufficiently large s ($s \geq \hat{s}(\varepsilon)$) $b_{rq}(x,\varepsilon,s) \sim O(\varepsilon^n)$, then the domains $\Omega^{(s)}$ will called weakly connected. Such domains exist only in spaces R^n of dimension $n \geq 3$.

EXAMPLE

Let $F^{(s)}$ be the union of round tubes with porous walls and axes of tubes form the cubic periodic net of a period $h^{(s)} = O(s^{-1})$. Suppose that outward diameter of tubes equal $d^{(s)} = d\, h^{(s)}$ $(d < 1)$, the thickness of their walls $\delta^{(s)} = O(1/s^{1+\gamma}) \leq \delta d^{(s)}$ $(\gamma \geq 0,\ \delta < 1/2)$ and radius of pores (cylindrical channels in walls) $r^{(s)} = O(1/s^{2+\gamma})$. If pores are distributed not very densely, then domains $\Omega^{(s)} = R^3 \setminus F^{(s)}$ do not satisfy the strong connectivity condition, they consist of the exterior $\Omega_1^{(s)}$ and interior $\Omega_2^{(s)}$ of the tubes, which are weakly connected via pores $Q^{(s)}$ because

$$b_{rq}(x,\varepsilon,s) \sim \frac{\pi\,(r^{(s)})^2\, n^{(s)}}{\delta^{(s)}\,(h^{(s)})^3} \cdot \operatorname{mes} K_\varepsilon^x = O(\varepsilon^3)$$

where $n^{(s)}$ - number of pores on cell of periods.

The following theorem describes the asymptotic behaviour of solutions $u^{(s)}(x)$ of Neumann problem in weakly connected domains $\Omega^{(s)}$ [3].

THEOREM 3

Let us assume, that for each subdomain $\Omega_r^{(s)}$ ($r = 1,2,..,m$) conditions (i) - (ii) of theorem 2 are valid, and $a_{ik}(x)$, $m^r(x)$ are corresponding limit functions. Let condition

$$\text{(iii)}\qquad \lim_{\varepsilon\to 0}\ \underline{\lim}_{s\to\infty} \frac{b_{rq}(x,\varepsilon,s)}{\operatorname{mes} K_\varepsilon^x} = \lim_{\varepsilon\to 0}\ \overline{\lim}_{s\to\infty} \frac{b_{rq}(x,\varepsilon,s)}{\operatorname{mes} K_\varepsilon^x} = b_{rq}(x)$$

also is valid.

Then the solution $u^{(s)}(x)$ of the problem (2.1) - - (2.2) for $s \to \infty$ converges in the following sense

$$\lim_{s\to\infty} \sum_{r=1}^{m} \| u^{(s)}(x) - u_r(x) \|^2 \chi_r^{(s)} dx = 0$$

to m-component function $u(x) = \{u_1(x), u_2(x), \ldots, u_m(x)\}$, which is the solution of following homogenized problem

$$(2.5) \quad \frac{1}{m_r(x)} \sum_{i,k=1}^{n} \frac{\partial}{\partial x_i} \left(a_{ik}^{r}(x) \frac{\partial u_r}{\partial x_k} \right) - \sum_{r,p=1}^{n} c_{rp}(x) u_p(x) +$$

$$+\lambda\, u_r(x) = f(x) \;, \; x \in \Omega$$

$$(2.6) \qquad \frac{\partial u_r}{\partial \nu_r} = 0, \; x \in \partial\Omega \; (r = 1,2,\ldots,m)$$

where $\frac{\partial}{\partial \nu_r}$ is conormal derivative ,corresponding to coefficients $a_{ik}(x)$.

Thus, the scalar Neumann problem (2.1) - (2.2) in weakly connected domains after homogenization gives the vector problem (2.5) - (2.6).

2.3. Domains with accumulators

Let us consider another type of domains $\Omega^{(s)}$, which do not satisfy the strong connectivity condition.

Suppose, that for any s , the domains $\Omega^{(s)}$ may be represented on such form

$$\Omega^{(s)} = \Omega_0^{(s)} \cup \left(\bigcup_{j=1}^{s} G_j^{(s)} \right) \cup Q^{(s)}$$

where sequence of subdomains $\{ \Omega_0^{(s)}, s = 1,2,\dots\}$ and $\{ G_j^{(s)}, s = 1,2,\dots\}$ satisfy the strong connectivity condition diam $G_j^{(s)} \to 0$ for $s \to \infty$; the subdomains $G_j^{(s)}$ $(j = 1,2,\dots,s)$ do not intersect and weakly connected with $\Omega_0^{(s)}$ via the set $Q^{(s)}$, which Lebesgue measure decrease to zero for $s \to \infty$. The quantitative characteristic of weak connection is introduced by mean of functional.

$$P_{x\varepsilon}^{(s)}[t^{(s)}] = \inf_{v^{(s)}} \int\limits_{\Omega^{(s)} \cap K_\varepsilon^x} \Big\{ |v^{(s)}|^2 + \varepsilon^{-2-\theta} \Big[|v^{(s)}|^2 \chi_0^{(s)} +$$

$$+ \sum_{j=1}^{s} |v^{(s)} - t_j^{(s)}|^2 \chi_j^{(s)} \Big\} dx = \sum_{i,j=1}^{s} p_{ij}(x,\varepsilon,s) t_i^{(s)} t_j^{(s)}$$

where $\chi_0^{(s)}(x)$, $\chi_j^{(s)}(x)$ are characteristic functions of subdomains $\Omega_0^{(s)}$ and $G_j^{(s)}$ $(j = 1,2,\dots,s)$ respectively.

Weak connection between subdomains $\Omega_0^{(s)}$ and $G_j^{(s)}$ in the ball $K_\varepsilon^x \supset G_j^{(s)}$ means, that $p_{ij}(x,\varepsilon,s) = O(\text{mes } K_\varepsilon^x \cap G_j^{(s)})$. We shall call such domains as domains with accumulators.

EXAMPLE

Let $F^{(s)}$ be the union of spherical shells with porous walls centered at cubic lattice in R^3 of the period $h^{(s)} = O(1/s)$. Suppose , that outward radius of shells $R^{(s)} =$

$= a\, h^{(s)} (a < 1/2)$, their thickness $\delta^{(s)} = O(1/s^{1+\gamma}) \leq \delta\, R^{(s)}$ $(\gamma \geq 0,\ \delta < 1)$ and each shell has $n^{(s)}$ cylindrical channels of radius $r^{(s)} = O(1/s^{2+\gamma})$. Then domains $\Omega^{(s)} = R^3 \setminus F^{(s)}$ do not satisfy the strong connectivity condition, but they are domains with accumulators because

$$p_{ij}(x,\varepsilon,s) \sim \frac{\pi\,(r^{(s)})^2\, n^{(s)}}{\delta^{(s)}} = O(\text{mes } K_\varepsilon^x \cap G_j^{(s)})$$

The following theorem describes the asymptotic behaviour of solutions $u^{(s)}(x)$ of Neumann problem in domains with accumulators [4].

<u>THEOREM 4</u>

Let us assume, that conditions (i) - (ii) of theorem 2 are valid for subdomains $\Omega_0^{(s)}$ and $G^{(s)} = \bigcup_{j=1}^{s} G_j^{(s)}$ ($a_{ik}^0(x)$, $m_0(x)$ and $m(x)$ are corresponding limit functions) and uniformly with respect to $t^{(s)} = \{ t_1^{(s)}, t_2^{(s)}, \ldots, t_s^{(s)}\} \in R^s$:

$$\text{(iii)} \quad \lim_{\varepsilon\to 0} \underline{\lim}_{s\to\infty} \Lambda_{x\varepsilon}^{(s)} [t^{(s)}] = \lim_{\varepsilon\to 0} \overline{\lim}_{s\to\infty} \Lambda_{x\varepsilon}^{(s)} [t^{(s)}] = p(x)$$

where

$$\Lambda_{x\varepsilon}^{(s)} [t^{(s)}] = \frac{\sum_{i,j=1}^{n} p_{ij}(x,\varepsilon,s)\, t_i^{(s)} t_j^{(s)}}{\sum_{i=1}^{n} (t_i^{(s)})^2 \text{mes}\left[G_j^{(s)} \cap K_\varepsilon^x\right]}$$

Then the solution $u^{(s)}(x)$ of the problem (2.3) - (2.4) for $s \to \infty$ converges in $L_2(\Omega_0^{(s)})$-norm to the solution u(x) of the following homogenized problem:

(2.7)
$$\frac{1}{m_0(x)}\sum_{i,k=1}^{n}\frac{\partial}{\partial x_i}\left(a^0_{ik}(x)\,\frac{\partial u}{\partial x_k}\right) - \lambda\left[1 + \frac{q(x)}{p(x)-\lambda}\right]u(x) = \left[1 + \frac{q(x)}{p(x)-\lambda}\right]f(x)\ ,\quad x \in \Omega$$

(2.8)
$$\frac{\partial u(x)}{\partial \nu_0} = 0\ ,\ x \in \partial\Omega$$

where $q(x) = p(x)\,m(x)/\,m_0(x)$.

Note, that the spectral parameter λ after homogenization enters into homogenized equation (2.7) nonusually, by mean of the rationals functions. This fact leads to homogenized models with memory in cases of non-stationary problems. We illustrate this by the example of diffusion equation.

Let us consider the following initial boundary-value problem:

(2.9)
$$\frac{\partial u^{(s)}}{\partial t}(x,t) - \Delta\, u^{(s)}(x,t) = f(x,t)\ ,\ x \in \Omega^{(s)},\ t > 0$$

(2.10)
$$\frac{\partial u^{(s)}}{\partial \nu}(x,t) = 0\ ,\ x \in \partial\Omega^{(s)},\ t > 0$$

$$u(x,0) = U(x), \; x \in \Omega^{(s)}$$

where $U(x) \in H^1(\Omega)$, $f(x,t) \in L_2(\Omega_T)$ $(\Omega_T = \Omega \times [0,T])$.

THEOREM 5

Let us assume that conditions (i) - (iii) of theorem 4 are valid .

Then the sequence of solutions $u^{(s)}(x,t)$ of the problem (2.9) - (2.10) for any $t \geq 0$ converges in $L_2(\Omega_0{}^{(s)})$ to the solution $u(x,t)$ of the following homogenized problem:

$$\frac{\partial u}{\partial t} - \frac{1}{m_0(x)}\sum_{i,k=1}^{n}\frac{\partial}{\partial x_i}\left(a^0_{ik}(x)\frac{\partial u}{\partial x_k}\right) + q(x)\Big[u - p(x)\int_0^t e^{-p(x)(t-\tau)} u(x,\tau)\, d\tau\Big] = f(x,t) + q(x)\int_0^t e^{-p(x)(t-\tau)} f(x,\tau)\, d\tau + q(x)\; U(x)\, e^{-p(x)t}, \qquad x \in \Omega_T,\; t > 0, \tag{2.11}$$

$$\frac{\partial u}{\partial \nu_0}(x,t) = 0,\; x \in \partial\Omega, t > 0;\; u(x,0) = U(x), x \in \Omega, \tag{2.12}$$

Note, that time-integral term appears in the homogenized equation (2.11) and this may be treated as homogenized model has 'memory'. Domains with accumulators arise in the theoty of non-stationary filtration of liquids and gases in so-called cracked-porous media [5], where the substance moves through cracks $\Omega_0^{(s)}$ and it accumulates in inner pores $G_j^{(s)}$ of blocks of the rock, which are weakly connected with cracks via boundary pores $Q^{(s)}$ of blocks. Media of such structure have memory.

Equations with rapidly oscillating coefficients.

Weakly connected domains and general notion of weak connection arise in the theory of highly temperature superconductivity, where composite media are superconductivity ceramics. Physical processes in such media are described by partial differential equations with rapidly oscillating coefficients which do not satisfy conditions of uniform ellipticity or boundedness. The homogenization of such equations leads to unusual homogenized models similar to above. We illustrate this for case parabolic equation of second order.

Let us consider in a domain $\Omega \subset R^n$ $(n \geq 2)$ initial boundary-value problem:

$$\frac{\partial u^{(s)}}{\partial t} - \sum_{i,k=1}^{n} \frac{\partial}{\partial x_i}\left(a_{ik}^{(s)}(x) \frac{\partial u}{\partial x_k}\right) = f(x,t) \qquad x \in \Omega,\ t > 0 \tag{3.1}$$

$$\frac{\partial u^{(s)}}{\partial \nu_0}(x,t) = 0,\ x \in \partial\Omega^{(s)},\ t > 0;$$

(3.2)

$$u^{(s)}(x,0) = U(x), \quad x \in \Omega$$

where $f(x,t) \in L_2(\Omega_T)$, $U(x) \in H^1(\Omega)$, $\frac{\partial}{\partial \nu}$ is conormal derivative on Ω, corresponding to coefficients $a_{ik}(x)$.

We assume, that coefficient $a_{ik}(x)$ depend on natural

parameter s and for any s they satisfy ellipticity and boundedness conditions:

$$(3.3)\qquad \alpha^{(s)}(x)\,|\xi|^2 \le \sum_{i,k=1}^{n} a_{ik}^{(s)}(x)\xi_i\xi_k \le \beta^{(s)}(x)\,|\xi|^2$$

$$(0 \le \alpha^{(s)}(x) \le \beta^{(s)}(x) < \infty,\ x \in \Omega)$$

It is known, that if condition (3.3) is fulfilled uniformly with respect to s, i.e. there exist constants α, β such that

$$0 \le \alpha \le \alpha^{(s)}(x) \le \beta^{(s)}(x) \le \beta < \infty, x \in \Omega$$

then the homogenized equation for the problem (3.1) - (3.2) has the same form as the initial equation (3.1).

$$\frac{\partial u}{\partial t} - \sum_{i,k=1}^{n} \frac{\partial}{\partial x_i}\left(a_{ik}(x) \frac{\partial u}{\partial x_k}\right) = f(x,t),\ x \in \Omega\, t > 0$$

It is simplest homogenized model of physical processes in homogeneous media (local scalar model). But, if the condition of uniform ellipticity or boundedness is not fulfilled, e.g. when there exist subsets $F^{(s)} \subset \Omega$ (s = 1,2,...) such that sup $\alpha^{(s)}(x) \to 0$ or inf $\beta^{(s)} \to \infty$ for $s \to \infty$, then homogenized equations are more complicated.Their form essentially depending on the geometric structure of the sets $F^{(s)}$.

Consider the most characteristic cases.

3.1. Non-local homogenized model.

Let us assume first that the condition of uniform ellipticity is fulfilled , but the condition of uniform boundedness is not fulfilled. Namely, let there exist constants α, β and subdomains $F^{(s)} \subset \Omega$, such that

(i) $\inf_{\Omega} \alpha^{(s)} \geq \alpha > 0, \quad \sup_{\Omega \setminus F^{(s)}} \beta^{(s)}(x) < \beta < \infty.$

$$\alpha^{(s)} = \inf_{F^{(s)}} \alpha^{(s)}(x) \to \infty \quad (s \to \infty)$$

(ii) $\operatorname{mes} F^{(s)} = o\left(\frac{1}{\alpha^{(s)}} \right) \to 0 \quad (s \to \infty) ;$

(iii) domains $F^{(s)}$ $(s = 1,2,\ldots)$ and $\Omega\setminus F^{(s)}$ $(s = 1,2,\ldots)$ satisfy the strong connectivity condition .

In this case, for s ∞ , the solution $u^{(s)}(x,t)$ of the problem (3.1) - (3.2) for any $t > 0$ converges in $L_2(\Omega)$ to the solution u(x,t) of following homogenized problem:

$$\frac{\partial u}{\partial t} - \sum_{i,k=1}^{n} \frac{\partial}{\partial x_i}\left(a_{ik}(x) \frac{\partial u}{\partial x_k}\right) + C(x)\, u -$$

$$- \int_{\Omega} R(x,y)u(y,t)dy = f(x), \; x \in \Omega, \; t > 0$$

$$\frac{\partial u}{\partial \nu_a} = 0, \; x \in \partial\Omega \;, \; t > 0; \quad u(x,0) = U(x), \; x \in \Omega,$$

where $R(x,y) = C(x)\,G(x,y)C(y)$, $G(x,y)$ is the Green function of the of the boundary-value problem:

$$\sum_{i,k=1}^{n} \frac{\partial}{\partial x_i}\left(b_{ik}(x) \frac{\partial G}{\partial x_k}\right) - C(x)G = -\delta(x,y) \quad , \; x\in \Omega, y\in\Omega;$$

$$\frac{\partial G}{\partial \nu_b}(x,y) = 0 \quad , \; x \in \partial\Omega;$$

The exact conditions of convergence and definition of the tensors $\{a_{ik}(x)\}_{ik}^{n}$, $\{ b_{ik}(x)\}_{ik}^{n}$ and function $C(x)$ are given in [6].

EXAMPLE

Let $F^{(s)}$ be a union of round cross section threads with axes forming in R^3 a periodic net of the period $h^{(s)}=O(1/s)$, and radius $r^{(s)}$. Let the coefficients of initial equation (3.1) be determined by the equalities

$$a_{ik}(x) = \begin{cases} \delta_{ik}, & x \in\Omega\setminus F^{(s)} \\ a^{(s)}\delta_{ik}, & x\in F^{(s)} \end{cases} \tag{3.5}$$

where $a^{(s)} = O((h^{(s)}/\,r^{(s)})^2) \to \infty \;(s\to \infty)$

Then the coefficients of homogenized equations are determined by following formula

$$a_{ik}(x) = \delta_{ik}; \qquad b_{ik}(x) = \lim_{s\to\infty} \frac{\pi (r^{(s)})^2 a^{(s)}}{(h^{(s)})^2}\,\delta_{ik},$$

$$C(x) = \lim_{s\to\infty} \frac{6\pi}{(h^{(s)})^2 |\ln r^{(s)}|}$$

3.2. Homogenized vector model

Let us assume now the condition of uniform boundedness is fulfilled, but the condition of uniform ellipticity is not fulfilled. Namely, there exist constants α, β and closed subsets $F^{(s)}$ such that

(i) $\sup_{\Omega\setminus F^{(s)}} \beta^{(s)}(x) < \beta < \infty$, $\inf \alpha^{(s)}(x) \geq \alpha \geq 0$,

$$\inf \alpha^{(s)}(x) \to 0 \quad (s \to \infty)$$

(ii) $\quad \operatorname{mes} F^{(s)} \to 0 \quad (s \to \infty)$

(iii) $\Omega = \Omega_1^{(s)} \cup \Omega_2^{(s)} \cup F^{(s)}$, where the sequences of domains $\{\Omega_1^{(s)}; s = 1,2,\ldots\}$ and $\{\Omega_2^{(s)}; s = 1,2,\ldots\}$ both satisfy the strong connectivity condition.

In this case, for $s \to \infty$, the solution $u^{(s)}(x,t)$ of the problem (3.1) - (3.2) for any $t > 0$ converges , in the following sense

$$\lim_{s\to\infty} \sum_{i=1}^{2} \|u^{(s)}(\cdot,t) - u_i(\cdot,t)\|_{L_2(\Omega_i^{(s)})} = 0$$

to the solution $(u_1(x,t), u_2(x,t))$ of such homogenized problem

$$\frac{\partial u_1}{\partial t} - \frac{1}{m_1(x)}\sum_{i,k=1}^{n}\frac{\partial}{\partial x_i}\left(a_{ik}^{1}(x)\frac{\partial u_1}{\partial x_k}\right) + C_1(x)(u_1 - u_2) =$$

$$= f_1(x,t)\ , \quad x \in \Omega,\ t > 0;$$

$$\frac{\partial u_2}{\partial t} - \frac{1}{m_2(x)}\sum_{i,k=1}^{n}\frac{\partial}{\partial x_i}\left(a_{ik}^{2}(x)\frac{\partial u_2}{\partial x_k}\right) + C_2(x)(u_2 - u_1) =$$

$$= f_2(x,t)\ , \quad x \in \Omega,\ t > 0$$

$$\frac{\partial u_r}{\partial \nu_r} = 0,\quad x \in \partial\Omega\ ,\quad t > 0;$$

$$u_r(x,0) = U(x),\quad x \in \Omega, (r = 1,2)$$

The conditions of convergence and the definition of functions $a_{ik}^{r}(x)$, $m_r(x)$ and $C_r(x)$ (r = 1,2)are given in Ref.7.

Note, that above geometrical structure of sets $F^{(s)}$ exist only in the space R^n of a dimension $n \geq 3$ .In a typical situation $F^{(s)}$ represents a highly branched system of thin-walled tubes and the domains $\Omega_1^{(s)}$ and $\Omega_2^{(s)}$ are interior and exterior of tubes respectively.

3.3. Homogenized model with memory

Let us assume that the conditions (i) ,(ii) of the section 3.2 are valid, but sets $F^{(s)}$ have another geometrical structure so that

(iii)$\Omega = \Omega_0^{(s)} \cup (\bigcup_{j=1}^{s} G_j^{(s)}) \cup F^{(s)}$.where the sequences of domains $\{\Omega_0^{(s)};\ s = 1,2,\dots\}, \{G_j^{(s)}; s = 1, 2, \dots\}$ $(j = 1,\dots,s)$ satisfy the strong connectivity condition.

In this case, for $s \to \infty$ the solution $u^{(s)}(x,t)$ of the problem (3.1) - (3.2) for any $t > 0$ converges in $L_2(\Omega_0^{(s)})$-norm to the solution of the following homogenized problem:

$$\frac{\partial u}{\partial t} - \frac{1}{m_0(x)} \sum_{i,k=1}^{n} \frac{\partial}{\partial x_i}\left(a_{ik}^0(x) \frac{\partial u}{\partial x_k}\right) +$$

$$+ p(x)\,(m_0^{-1}(x) - 1) \int_0^t e^{-p(x)(t-\tau)} \frac{\partial u}{\partial \tau}\, d\tau = f(x,t) +$$

$$+ p(x)\,(m_0^{-1}(x) - 1) \int_0^t e^{-p(x)(t-\tau)} f(x,\tau)\, d\tau\ ,\ x \in \Omega,\ t > 0$$

$$\frac{\partial u}{\partial \nu_0}(x,t) = 0,\ x \in \partial\Omega,\ t > 0; \quad u(x,0) = U(x),\quad x \in \Omega,$$

The conditions of convergence and the definitions of the tensor $\{a_{ik}(x)\}_{ik}^{n}$ and functions $m_0(x)$ and $p(x)$ are given in Ref.8.

EXAMPLE

Let $F^{(s)}$ be the union of spherical shells of the

radius $r^{(s)} = r/s$ and thickness $\delta = o(r^{(s)})$, the centres of shells form a cubic lattice in R^3 of the period $h^{(s)} = h/s$. Let the coefficient of initial equation (3.1) be determined by the equation (3.5), where $a^{(s)} = O(\delta^{(s)} h^{(s)})$, $a^{(s)} \to 0$ for $s \to \infty$.

Then the coefficients of homogenized equations are determined by formulae

$$m_0(x) = 1 - \lim_{s\to\infty} \frac{4\pi (r^{(s)})^3}{3\,(h^{(s)})^3} \; ; \quad p(x) = \lim_{s\to\infty} \frac{3\, a^{(s)}}{\delta^{(s)} r^{(s)}};$$

$$a^0_{ik}(x) = \delta_{ik} \int_{K \setminus Q} |\nabla v|^2 dx,$$

where v (x) is the solution of the cell problem:

$$\Delta v(x) = 0 , \; x \in K \setminus \Omega; \quad \frac{\partial v}{\partial \nu}(x) = 0 \;\; x \in \partial Q \cup \partial K \setminus (\Gamma_+ \cup \Gamma_-) ; \; v(x) = \pm 1/2 , \; x \in \Gamma_\pm$$

in the unit cube K with the lacking central sphere Q of the radius r/h; $\Gamma_\pm$ are two opposite faces of the cube K

References

[1] V.A.Marchenko and E.Ya.Khruslov,"Boundary-value problems in domains with fine-grained boundaries", Naukova Dumka, Kiev, 1974,278p.

[2] E.Ya.Khruslov, "The asymptotic behaviour of the solution of the second boundary-value problem for diminution of grains of the boundary", Mat, Sbornik 106 No 4 (1978)

[3] E.Ya.Khruslov,"On convergence of the solution of the

second boundary-value problem in weakly connected domains" ,In: "Theory of operators in functional spaces and its applications"(in Rus.), Naukova Dumka, Kiev, 1981.

[4] E.Ya.Khruslov, "Homogenized models of diffusion in cracked-porous media" ,Dokl. AN SSSR, 309 No 2, 1989

[5] G.I.Barenblat, V.M.Entov and V.M.Ryzhik,"Motion of liquids and gases in natural layers"(in Rus.), Nauka, Moscow, (1984) ,208p.

[6] V.N.Fenchenko and E.Ya.Khruslov, "Asymptotic of of solution of differential equations the with strongly oscillating matrix of coefficients which which does not satisfy the condition of uniform boundedness", Dokl. AN Ukr.SSR, No. 4,(1981).

[7] V.N.Fenchenko and E.Ya.Khruslov, "Asymptotic of solution of differential equations with strongly oscillating and degenerating matrix of coefficients", Dokl. AN Ukr.SSR, No.4,(1980).

[8] E.Ya.Khruslov, "A homogenized model of highly inhomogeneous medium with memory", Uspekhi Matem.Nauk. (1990)

E.Ya. Khruslov
Ukrainian SSR Academy of Sciences
Physico-Technical Institute of Low Temperatures
Kharkov
USSR

Structural Optimization of a Linearly Elastic Structure using the Homogenization Method

Noboru Kikuchi and Katsuyuki Suzuki

Abstract

We shall describe a brief review of the structural optimization of a linearly elastic structure, and we shall present a new method to solve the sizing, shape, and layout (topology) problems based on the theory of homogenization. Many numerical examples of the optimal design are also presented as well as a mathematical formulation of a relaxed design problem.

1. Introduction

There are three major structural optimization problems of a linearly elastic structure ; namely, 1) sizing, 2) shape, and 3) layout(topology) optimization problems. The characteristics of these problems can be summarized as follows :

<u>Sizing Problem</u> A typical setting of the problem is to find the optimal thickness distribution of a linearly elastic plate that is supported on its boundary and is subject to a given loading condition on the plate or its boundary. The optimal thickness of the plate is obtained so as to, e.g., minimize (or maximize) a certain physical quantity such as the mean compliance, while the state (i.e. equilibrium) equation of the structure is maintained as well as various constraints on the state and design variables. Here, the state variable can be the deflection of the plate, and the thickness of the plate becomes the design variable. If a beam is subject to axial torsional and bending forces, the sizes of the cross section of the beam such as the radius, height, and width would be optimized to construct the strongest beam structure within a given design restriction. That is, the design variable of the sizing problem is a physical dimension of the structure.

The main feature of the sizing problem is that the domain of the design and state variables is *a priori* known and is fixed in the optimization process. For example, the optimum thickness distribution is considered over the whole plate, the domain of the thickness and deflection is the whole middle surface Ω of the plate that is not altered by design optimization.

Shape Problem On the other hand, the shape problem is defined on a domain A which is unknown *a priori*. For certain restricted problems, the shape of the domain A may be described parametrically on a fixed parametric space, but the state equation must be defined on a "variable" domain A. The optimal shape of the domain, that is, the optimal domain A is obtained so as to minimize e.g. the mean compliance of the structure, while the equilibrium (i.e. state) equation is satisfied on such a domain A.

Layout (Topology) Problem Ambiguity of the layout (topology) problem is substantial in its mathematical description. In this case, only the known quantities are applied loads, desired support conditions, various design restrictions, and the possibly the volume of a structure constructed. The purpose of optimization is to find the optimal layout of a structure so that given applied loads are transmitted to desired supports in a specified region using a restricted amount of material while equilibrium and design constraints are satisfied. In this problem, not only the physical size but also the shape of the structure are unknown, even we cannot define these geometrical quantities ! In order to define these, the topology of the domain (structure) must be known, but it is not known *a priori*.

A schematic description of sizing shape and layout problems is given in Figure 1 for intuitive grasp of problems in structural optimization.

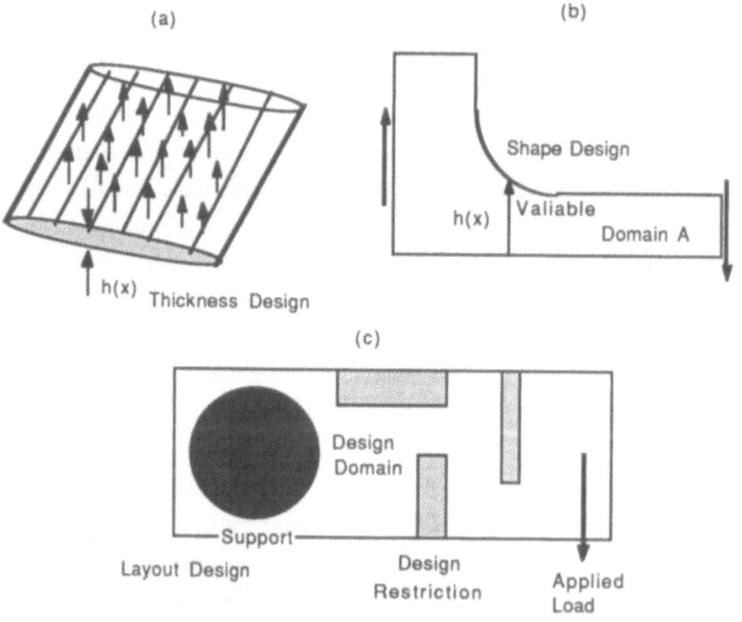

Figure 1. (a) Sizing, (b) Shape, (c) Layout Optimization Problem

the same with V except $u_{gi} = 0$, $i=1$ and 2. Young's modulus and Poisson's ratio are expressed by E and ν, respectively. If the "cost" function f is defined as the mean compliance of the elastic structure, this is given by

$$f(d,u) = L(u) + \int_{\partial A_d} \sum_{i,j=1}^{2} n_j \sigma_{ji}(u) u_{gi} d\Gamma \tag{6}$$

where $\sigma_{ji}(u)$ is the (j,i) component of the stress tensor at the equilibrium u. If there is no additional constraint on the displacement, $g_u = 0$. If the shape optimization is restricted to be within the "volume" of material Ω_0, then the constraint on the design variables $d = \{ d_1, d_2, ..., d_n \}$ is specified so as to $A \le \Omega_0$. Now, the shape optimization is to find the optimum design d which is *finite* number of parameters that defines the shape of the domain $h(x)$. If the basis functions $h_i(x)$ are chosen appropriately, this problem can be solved rather easily by applying a mathematical programing method, since the total number of design variables is finite.

3. Difficulties in Structural Optimization

If the structural optimization problem can be represented by using finite number of design parameters together with well-behaved basis functions, we do not have any difficulties both in the mathematical theory and in the computational method of optimization. It may be rather simple to justify the problem mathematically using a standard convergence argument to show existence of an optimal solution if it is represented by a fixed finite number of parameters, while it is much simpler to find an appropriate computational method to obtain the optimal solution numerically. Indeed, applying the so-called design sensitivity analysis, we can find the linearized problem which can be solved, e.g., by any linear programing methods. However, the problem encountered in engineering practice is that *the design problem cannot be represented by finite number of design parameters with well-behaved basis functions* ! We shall explain this feature using various examples.

A typical problem in the sizing problem is to determine the optimal thickness distribution of a linearly elastic plate that minimizes the mean compliance, or that maximizes the first eigenvalue of the free vibration problem of a plate. If the thickness distribution is represented by a fixed finite number of C^∞- basis functions, then the optimal solution can be easily obtained for such a choice as shown in, e.g., Banichuk[1]. However, Cheng and Olhoff[2] show that if a different selection of a finite number of basis functions is made, the optimal thickness distribution becomes so different from the others. This suggests that the optimal thickness distribution of all the possible choice cannot be represented as a limit of the optimal solutions of a fixed finite number of design parameters with basis functions. In engineering practice it is known that discrete rib reinforcement, see Figure 2, can produce a stiff plate. Thus, it is expected that the optimal thickness $h(x)$ is merely a function of $L^\infty(\Omega)$ or possibly a function of $H^{1/2-\delta}(\Omega)$ for $\delta>0$ at most. Note that infinitely many C^∞ functions are required to represent such a discrete rib.

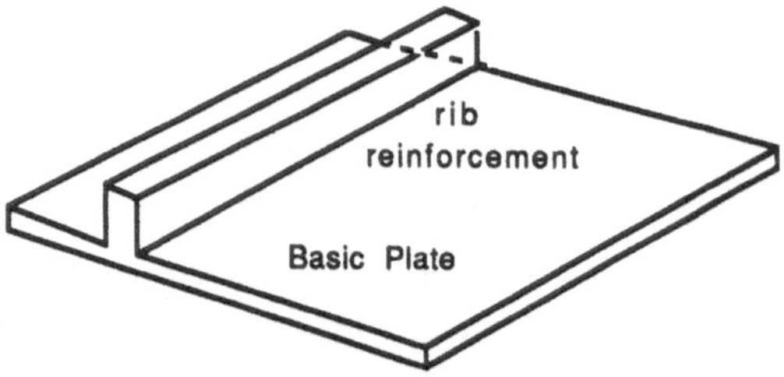

Figure 2. Discrete Reinforcement of a Plate

From the examination of the generalized state(equilibrium) equation we can expect convergence of $h^n = \text{span}\{h_1, ..., h_n\}$ at most in $L^\infty(\Omega)$ as n goes to ∞. This is implied by the fact that the coefficient tensor of the differential operator need be merely in $L^\infty(\Omega)$ in the generalized state equation in the context of variational principle, and then the assumption of the isotropic plate could be contradicted. That is, we have to state the equilibrium equation in a "relaxed" form using the assumption of an orthotropic plate that possibly possesses variety of microstructures instead of restricting our attention only for too well-behaved isotropic plate that assumes smooth variation of thickness as well as Young's modulus and Poisson's ratio. Lurie, Fedorov, and Cherkaev[3] studied this nature mathematically using the theory of homogenization, and provided a clear answer how this problem should be solved, see also Bendsøe[4].

In the shape problem difficulty we can find is not only for convergence of the parametric representation of the shape, but also for representation itself. For example, how can we represent the domain shown in Figure 3 parametrically with well-behaved basis function $h_i(x)$? Furthermore, there is no assurance that the singly connected domain is optimal. In fact, we know that there should be holes in a structure from our experience. But, we do not know the number of holes and their location *a priori*. Can we explicitly represent this situation parametrically with a finite number of definite basis functions ? The answer is very negative. Thus, the shape problem can only be solved with many additional restriction. For example, we restrict our attention just to a sub-optimal problem with the assumption of "manageable" parametric representation of a singly connected domain. Even within this restriction, we may have the same convergence problem with the sizing problem as the number of parameters goes to ∞! In other words, if a different parametric representation of the shape is assumed, the

2. Mathematical Formulation

A formal description of the structural optimization problem may be defined by minimizing (or maximizing) a fuctional with respect to the design variables consisting of {size, shape, topology} subject to {the state equation, constraints on the state variables, constraints on the design variables}. A typical mathematical setting is given by

$$\begin{aligned} &\underset{d}{Minimize} \qquad f(d,u) \\ &\text{subject to} \\ &u \in V : a(u,v) = L(v),\ \forall v \in V_0 \\ &g_u(u) \leq g_{u\max} \\ &g_d(d) \leq g_{d\max} \end{aligned} \tag{1}$$

Here, $f(d,u)$ is the "cost" function, d and u are the design and state variables, respectively, "$a(u,v)=L(v)\ \forall v$ " is a generalized (weak) form of the state equation, and g_u and g_d represent the constraints on the state and design variables, respectively.

As a concrete example of the structural optimization, we shall define a shape problem shown in Figure 1(b), that is to determine the optimal shape $h(x)$ which is parametrically represented by a set of basis functions { $h_1(x), h_2(x), \ldots, h_n(x)$ } , i.e.,

$$h(x) = \sum_{i=1}^{n} d_i h_i(x) \tag{2}$$

If the plane stress condition is assumed with material isotropy, the generalized form of the state (equilibrium) equation is given by

$$a(u,v) = \int_A \begin{Bmatrix} \varepsilon_1(v) \\ \varepsilon_2(v) \\ 2\varepsilon_{12}(v) \end{Bmatrix}^T \frac{E}{1-\nu^2} \begin{bmatrix} 1 & \nu & 0 \\ \nu & 1 & 0 \\ 0 & 0 & (1-\nu)/2 \end{bmatrix} \begin{Bmatrix} \varepsilon_1(u) \\ \varepsilon_2(u) \\ 2\varepsilon_{12}(u) \end{Bmatrix} d\Omega \tag{3}$$

$$L(v) = \int_A v^T b\, d\Omega + \int_{\partial A_t} v^T t\, d\Gamma \tag{4}$$

$$V = \left\{ v = \{v_1, v_2\} : v_i \in H^1(A),\ \gamma(v_i)\big|_{\partial A_d} = u_{gi}, u_{gi} \in H^{1/2}(\partial A_d) \right\} \tag{5}$$

$\varepsilon_{ij}(v) = \frac{1}{2}\left(\frac{\partial v_i}{\partial v_j} + \frac{\partial u_j}{\partial x_i}\right)$, b and t are the applied body force and traction on the part of the boundary ∂A_t, $H^1(A)$ is the Sobolev space defined on A, γ is the trace operator from $H^1(A)$ to $H^{1/2}(\partial A)$, u_{gi} is the prescribed displacement on the part of the boundary ∂A_d which is the complement of ∂A_t of the boundary ∂A of the domain A, and V_0 is

solution may be very different from the others. Suppose that we have a convergent optimal domain.

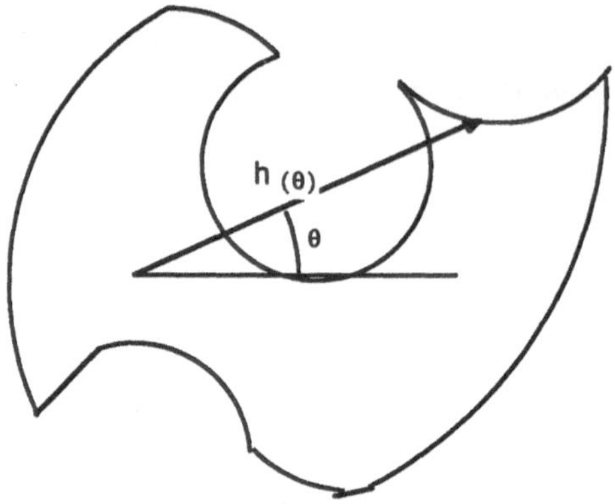

Figure 3. Complexity of the Shape Definition

Even with this assumption, there still exists a considerably difficult problem in the shape problem when it is solved numerically. Because of a finite number of parametric representation of the shape, the sensitivity analysis to obtained a linearized problem is rather straightforward. But, solving the generalized state equation using any kind of discretization methods is not obvious. For example, the finite element method is applied, we have to triangularize the domain A by finite elements. Note that the initial shape of the domain is considerably different from the "current" one, see Figure 3. This means that discretization of the domain becomes a significant task, since it must be done automatically without any manual help nor terminating computation for optimization. Since the design variable is defined on a fixed domain, this discretization difficulty cannot be encountered in the sizing problem. The difficulty in discretization is a very special nature of the shape problem arose in numerical computation of the state variable. An automatic finite element mesh generation must be introduced to solve this in practice. As shown in above, there are many difficulties in the shape problem we must solve even for a very restricted case.

It is needless to say that what kind of difficulties we can find in the layout (topology) problem, since it is a complex combination of the sizing and shape problems if the topology of a structure is fixed with a finite number of parameters. Again it is not known whether or not the optimal topology can be represented by a finite number of parameters. Only available approach is to set up an initial set of truss joints and to generate additional ones within the limited total volume of the structure using, e.g. dynamic programing methods, see Palmer[5]. In this case, the type of structure must be predetermined, for example, as a truss structure that cannot transmit the bending moment at joints. But, this choice need not be the best. Furthermore, we can deal with only a fixed finite number of joints. There is, again, no assurance that a convergent optimal structure can be obtained as the number of joints is increased.

4. A New Approach Based on the Homogenization

As shown in above, there are many difficulties in structural optimization. Now we shall look at the issue from the opposite point of view, that is, we shall describe a desirable method to solve structural optimization problems, despite that its existence is in question.

A common difficulty in the sizing, shape, and layout problems is convergence of a finite number of parametric representation. This means that once a method is introduced to solve the convergence problem for one of these problems, it may be applicable for the others. There is a possibility of existence of a method to solve these three problems at once. To solve the shape problem, a common approach is to define the shape using certain functions and to vary it to achieve the optimality. Variation of the shape generates substantial difficulty in computation. Thus, we should abandon this approach. That is, the shape should not be defined by a set of functions, and the problem must be solved in a fixed domain as for the sizing problem so that finite element(difference) discretization need not be renewed during optimization. Furthermore, since the topology is unknown *a priori*, it should not be specified. From the very beginning, "infinite" number of holes should be prepared at "everywhere" to solve both the shape and topology problems.

Does there exist such a method ? To answer to this question, recall that the convergence problem in the sizing problem is solved by applying the concept of the homogenization by Lurie, Fedorov, and Cherkaev[3], and their method is applied to solve the thickness optimization problem by e.g. Bendsøe[4]. But this is bounded in the sizing problem.

Now recall the paper by Murat and Tartar[6] presented in 1983 on optimality conditions and homogenization. In this paper, Murat and Tartar stated "Where the variable is a domain, the computing of variations was done as early as 1950 by Hadamard, by pushing the boundary along the normal and then computing the induced variation of the functional. Turning this idea into theorems is not easy, and moreover already one can see another defect of this method : a given domain is compared only with domains of the same shape ; it is impossible to make a hole inside the domain or to add a few small pieces far away by means of this technique. ..... The real difficulty lies in the fact that the set of domains, i.e., characteristic functions, does not possessess natural paths from one domain to another : there is no manifold structure that enables us to use classical derivatives. ..... In the problem where the variable appears as a domain and some partial differential equation is involved, there is another phenomenon that we discovered ten years ago (it was later called homogenization) : generalized domains appear which are the analogue of a mixture of two different materials and the effective properties of these mixtures have to be understood (they are not obtained by averaging certain quantities in more than one dimension)."

The very similar but direct idea for structural optimization is stated by Kohn and Strang[7] : "The need for relaxation is reflected in the design problem by the possibility that there may be no optimal design. Though initially surprizing, this phenomenon is easily understood : it is sometimes advantageous to perforated parts of Ω by many fine holes with a suitably chosen geometry. If the optimal characteristics can be realized only in the limit as the scale of the perforation tends to zero, then the design problem has no geometrical solution - rather, a solution exists only in a suitably generalized class of designs, which allows the use of composites obtained by perforation at some points of Ω. ..... We first extend the design problem, allowing for the use of composite materials. Then we choose the best composite at each point."

It should be now clear that there is a method to solve quite wide range structural optimization problems using the notion of homogenization after extending the design problem by allowing possibly perforated materials for certain points of the structural domain. Since perforation is expected in the microstructure, and since the degree of perforation as well as location is not specified *a priori*, arbitrary shape and topology of the structure may be well represented if they exist. It is certain that we may end up a perforated structure as the optimum because of the relaxation (i.e. extension) of the original design problem that is restricted to the use of only a solid nonperforated material. Because of the convergence property in the sense of homogenization, a precise mathematical theory should be able to furnish to the relaxed i.e. extended problem of structural optimization. In short, the theory of the homogenization is the answer to the optimal design problem in structural analysis.

5. Relaxed Optimal Design Problem

We shall describe the relaxed design problem using microscale rectangular holes to perforate a structure, see Bendsøe and Kikuchi[8]. Suppose that the total volume of microscale holes is specified in a given design domain Ω, that is, the volume Ω_S of "solid" material distributed in the design domain is specified. For simplicity, the design domain is plane so that plane stress analysis is sufficient to compute displacements and stresses, while the shape of microscale voids is assumed to be rectangular as shown in Figure 4.

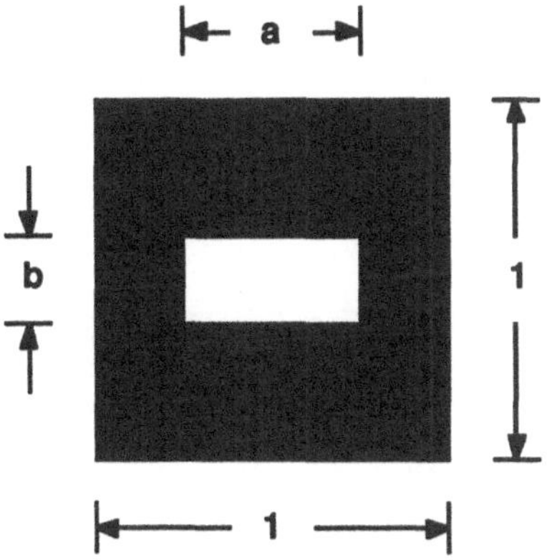

Figure 4. A Unit Cell Describing the Microstructure with a Rectangular Hole

Rectangular holes are chosen because they can realize the complete void ($a=b=1$) and solid ($a=b=0$) as well as generalized perforated medium ($0<a<1, 0<b<1$). If circular holes are assumed, they cannot reach to the complete void, and then are not appropriate to our purpose. It is noted that there are other choices to represent microscale holes such as using a generalize ellipse defined by $(y_1/a)^n+(y_2/b)^n = 1$, where a and b are the principal radii of the ellipse and n is the power defining the shape. However, in order to develop the complete void in the unit cell both a and b must be 1 as well as n goes to infinity for the generalize ellipse, while only a and b need be 1 if the rectangular hole is applied. Thus, using rectangular holes is simpler. Since holes are rectangular in the unit cell that characterize the microstructure of a perforated medium for the design problem, their orientation is important in the macroscopic problem for stress analysis. Indeed the anisotropic elasticity tensor in the macroscopic problem strongly depends on the orientation of microscale holes. Thus, the sizes a and b and the orientation θ of microscale rectangular holes must be the design variables of the relaxed problem.

Suppose that a, b, and θ are functions of the position x of an arbitrary point of a macroscale domain of a linearly elastic perforated structure Ω in the two-dimensional Euclidean space $\mathbb{R}^2$: $a=a(x)$, $b=b(x)$, and $\theta=\theta(x)$. Functions a, b, and θ may not be so smooth, i.e., it could be true that $a \in L^\infty(\Omega)$, $b \in L^\infty(\Omega)$, and θ maintains smoothness equivalent to the angle θ_p of the principal coordinates of the stress tensor $\sigma(x)$, i.e., θ may be continuous in Ω except a finite number of points. However, we shall assume that they are sufficiently smooth, for example, $a,b,\theta \in H^1(\Omega)$. Assuming that a periodic microstructure characterized by $a(x)$, $b(x)$, and $\theta(x)$ exists in a small neighborhood of an arbitrary point x in Ω, and assuming that such microstructure at x

need not be the same with the one at a different point x^*, see Figure 5 which shows a schematic setting of varying microstructures, a homogenized elasticity tensor $\mathbb{E}^H(x)$ is computed in order to solve a macroscopic stress analysis problem of a perforated structure.

Figure 5. Assumption of "Continuous" Change of Microstructures

The homogenized elasticity tensor is computed by solving the problem defined in the unit cell in which a rectangular hole is placed :

Find the characteristic deformations $\chi^{(kl)}=\{\chi_1^{(kl)},\chi_2^{(kl)}\}\in V_Y$ satisfying

$$\sum_{i,j,m,n=1}^{2}\int_Y E_{ijmn}\frac{\partial \chi_m^{(kl)}}{\partial y_n}\frac{\partial v_i}{\partial y_j}dY=\sum_{i,j=1}^{2}\int_Y E_{ijkl}\frac{\partial v_i}{\partial y_j}dY \quad ,\forall v\in V_Y \tag{7}$$

where $V_Y=\{v=\{v_1,v_2\} : v_i\in H^1(Y)\ ,\ v_i$ is Y-periodic in the unit cell Y, $i=1,2\}$, and the unit cell $Y=(-1/2,1/2)\times(-1/2,1/2)$. The elasticity tensor $\mathbb{E}$ is chosen either for the plane stress or for plane strain problem depending on the macroscopic perforated structure to be designed. The elasticity tensor $\mathbb{E}$ is zero if y is located in the hole, and coincides with the one of the "solid" material that is utilized to form a structure if y is outside of the hole. It is noted that Young's modulus does not affect to the relaxed design problem while Poisson's ratio may imply change to the optimal design. After obtaining the characteristic deformations $\chi^{(kl)}$, the homogenized elasticity tensor $\mathbb{E}^H$ is computed by

$$E_{ijkl}^{H}=\sum_{m,n=1}^{2}\int_{Y}\left(E_{ijkl}-E_{ijmn}\frac{\partial \chi_{m}^{(kl)}}{\partial y_{n}}\right)dY \tag{8}$$

Since the sizes $\{a,b\}$ of rectangular holes are functions of the position x, the homogenized elasticity tensor $\mathbb{E}^{H}$ varies in Ω. This means that the characteristic deformations must be obtained everywhere in the design domain Ω. Solving the unit cell problem (7) at everywhere is unrealistic. Thus, we shall solve (7) for several sampling points $\{a_i, b_j : i,j=1,\ldots,n\}$ of the sizes $\{a,b\}$ of rectangular holes, where $0 \le a_i \le 1$ and $0 \le b_j \le 1$, and we shall form a function $\mathbb{E}^{H} = \mathbb{E}^{H}(a,b)$ by an appropriate interpolation. The last step of obtaining the elasticity tensor for stress analysis of the macroscopic perforated structure is rotation of $\mathbb{E}^{H}$ by the angle θ. Defining the rotation matrix $\boldsymbol{R}$ by

$$\boldsymbol{R}(\theta)=\begin{bmatrix}\cos\theta & -\sin\theta\\ \sin\theta & \cos\theta\end{bmatrix}$$

the elasticity tensor $\mathbb{E}^{G}$ for stress analysis is computed by

$$E_{ijkl}^{G}(x)=\sum_{I,J,K,L=1}^{2} E_{IJKL}^{H}(a(x),b(x))R_{iI}(\theta(x))R_{jJ}(\theta(x))R_{kK}(\theta(x))R_{lL}(\theta(x)) \tag{9}$$

for $i,j,k,l = 1$, and 2, at arbitrary point x in Ω. It is clear that $\mathbb{E}^{G}$ is a function of the sizes $\{a,b\}$ and the rotation θ of microscale rectangular holes.

The state equation for stress analysis of the perforated structure is defined as follows, see Section 2 :

Solve the displacement $\mathbf{u}=\{u_1, u_2\}\in V$ satisfying

$$\sum_{i,j,k,l=1}^{2}\int_{\Omega}E_{ijkl}^{G}(x)\frac{\partial u_k}{\partial x_l}\frac{\partial v_i}{\partial x_j}d\Omega = L(v) \quad \forall v\in V_0 \tag{10}$$

Now, let us define an optimization problem using the simplest objective function, the mean compliance of the structure defined by (6), without any other side constraints on the state variable $\boldsymbol{u}$:

$$\begin{aligned}&\underset{a,b,\text{and }\theta}{Minimize}\quad f(d,u)\\&\text{subject to}\\&\quad(10)\\&\quad\text{and}\\&\int_{\Omega}(1-ab)\,d\Omega\le\Omega_s\end{aligned}\tag{11}$$

Here u is the solution of the principle of virtual displacement (10), $d=\{a,b,\theta\}$, and Ω_S is the total volume of "solid" material forming the "porous" structure. In general, Ω_S is much smaller than Ω that is the domain of the structure containing the design domain Ω, i.e., $\Omega_S < \Omega$.

Details of computational procedure including an optimization method to solve the design problem (11) can be found in Suzuki and Kikuchi[9], and are not discussed here.

6. Verification of the Relaxed Design Problem

We shall examine the present method to solve a simple structural design problem that the exact solution can be obtained analytically using a simple structural model such as a truss or a beam. A two-bar framed structure shown in Figure 6 is considered.

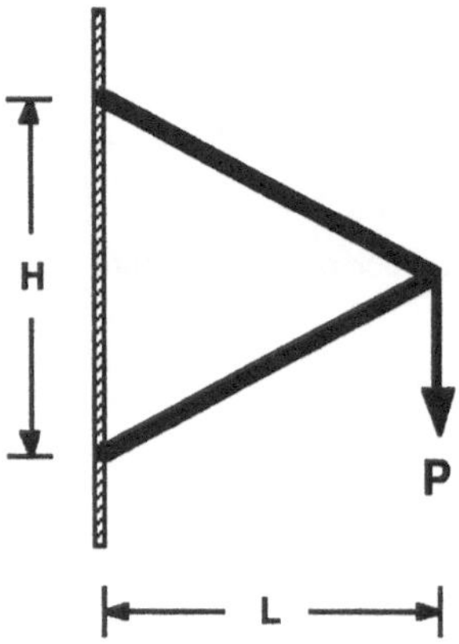

Figure 6. A Two-Bar Truss

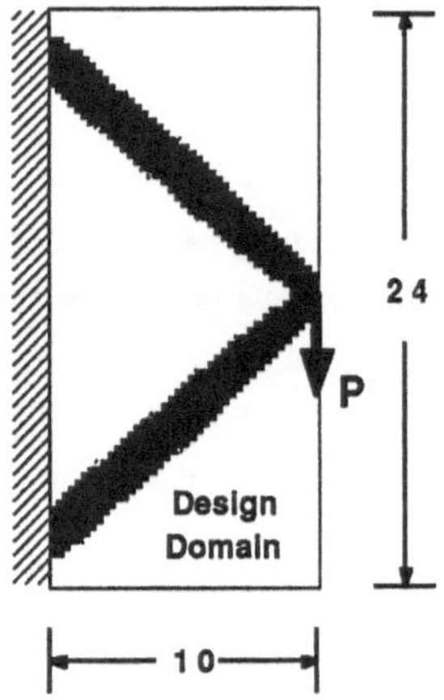

Figure 7. Optimal Relaxed Design
(Volume = 40 cm^2)

For the fixed values of the applied load P and the horizontal length L of the frames, the optimal height H is determined by minimizing the mean compliance. If the cross section of the truss-bars is unchanged to be rectangular with the unit width, the optimal height H is obtained as H=2L. Here the total volume of the frames is assumed to be constant. Defining a rectangular design domain Ω that is larger than the size Lx2L, where L=10cm, the relaxed design problem (11) is solved by discretizing Ω using 40 x 96 uniformly divided rectangular finite elements. The homogenized elasticity tensor $\mathbb{E}^H$ applied here is computed at 6 x 6 sampling points in the design variables *a* and *b*, respectively, and is interpolated by the Legendre polynomials. The homogenization problem (7) is solved using 10 x 10 finite elements for the "solid" isotropic material characterized by Young's modulus E=100GPa and Poisson's ratio ν=0.3. As shown in Figure 7, the present method forms a two-bar framed structure as the amount of the "solid" material is decreased, and the solutions converge to the analytical one obtained by assuming a simple truss structure. It is noted that the present method can provide not only the basic two-bar framed structure but also the size of the cross section of bars, especially for the case of large volume of the "solid" material.

7. Convergence of the Finite Element Approximation

The next verification issue is whether the shape and topology, i.e., the configuration of the structure obtained as the optimum in the relaxed design problem converges to the unique design as finite element meshes are uniformly refined, while other conditions are fixed. Importance of this test can be recognized by the observation in Cheng and Olhoff[2] that shows if the relaxed approach is not applied, the optimal rib distribution is highly mesh dependent. That is, refinement of the discretization leads completely different optimal solutions.

To check convergence property in discretization, let us solve the relaxed optimization problem (11) for a short cantilever subject to the vertical force at the free end, see Figure 8. For the volume 60 of the "solid" material, (11) is solved by using 32x20, 48x30, 64x40, and 80x50 equal size uniformly divided rectangular finite elements covering a rectangular design domain Ω. Applying the same homogenized elasticity tensor to the first example, the optimal configurations are obtained as shown in Figure 9. It is clear that the optimal configurations are convergent as the size of finite elements is reduced. Even every coarse meshes can provide sufficient idea of the topology and shape of the optimal structure. That is, if the present approach is used to find rough idea of the optimal structure, it is not necessary to solve the problem using very refined meshes. Figure 10 presents convergence of the mean compliance as the size of finite element goes to zero. Monotonic decreasing of the mean compliance is clearly observed.

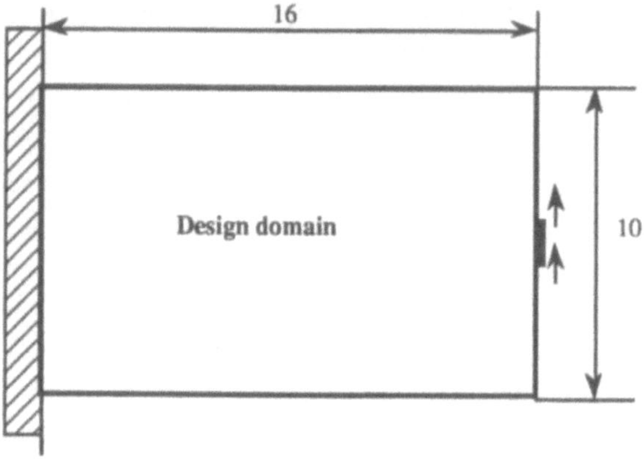

Figure 8. Design Domain for Bending of a Short Cantilever

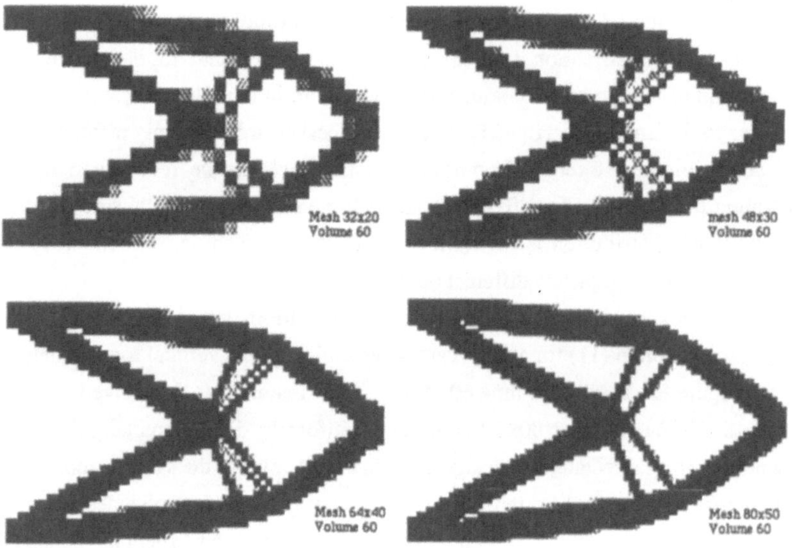

Figure 9. Convergence of the Optimal Configuration ($\Omega_S = 60$)

A very truss-like framed structure are built if the amount of the solid material is considerably smaller than that of the design domain. Each member is even straight. If the amount of the solid material becomes large, the optimal design diverges from a truss-like structure. Curved frames are generated, and more continuum-like shapes are formed. In this problem, despite of the relaxation that allows perforated composite at everywhere of the design domain, the optimal structure is not perforated at all. In other words, the relaxed problem provides the "classical" optimal solution to the design problem using only solid structural members.

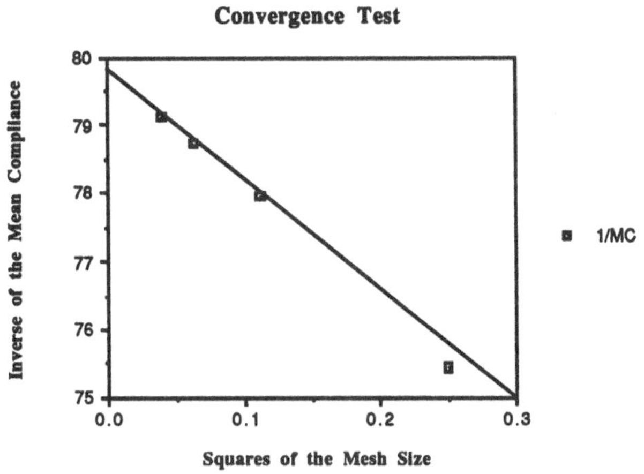

Figure 10. Convergence of the Mean Compliance

It is noted that from our other computational experience the perforated composite appears only in the vicinity of the portion where distributed traction or body forces are applied. If a finite number of discrete "point" forces is loaded, the optimal structure by the relaxed problem tends to generate a discrete solid structure rather than perforated composite. In the present example, only a point load is assumed at the free end, the optimal structure is a very discrete solid structure without any perforation. Instead of a distributed load applied along a portion of the boundary, if the displacement is specified over there, perforation does not also occur in the visinity of such boundary.

8. Design of a Framed Structure

We shall present an example how an optimal light weight structure is formed to minimize the mean compliance in a given load and support conditions by varying the volume of the solid material. To this end, let us consider a truss construction problem of a bridge as shown in Figure 11. The structure must be simply supported at the bottom end points, while a point load is applied at the center of the bottom line. As a design domain, a rectangle is assumed so that the height of the structure is restricted to be the half of the length L of the constructed bridge.

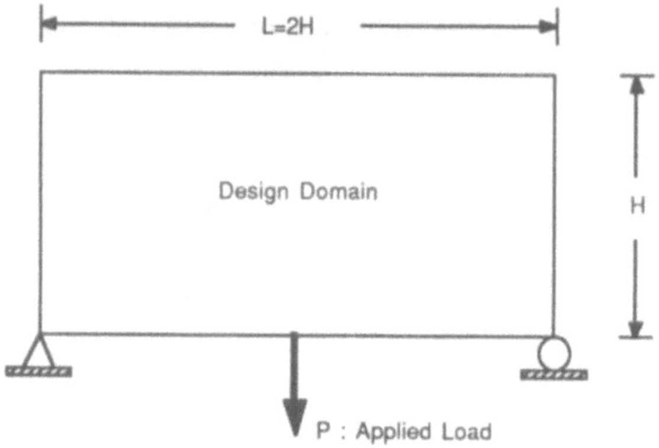

Figure 11. A Rectangular Design Domain for a Bridge Truss Design

For three different volumes of the solid material, the relaxed design problem (11) is solved, and the optimal solutions are obtained as shown in Figure 12. It is very clear that as the volume of the solid material is reduced, the structure becomes more truss-like, and that the optimal topology is easily identified. Four members connected at the loading point are straight and are subject to uniaxial tension forces, while two bars connected to the supports are subject to bending moments due to the applied load. Since these two bars are, roughly speaking, "simply" supported, higher stress distribution is expected at the center portion. In order to achieve uniform stress distribution in the bars, that is actually optimum for such truss structures, the bending rigidity must be large in the center portion. To this end, a long narrow slit(void) is formed over there. The optimal solution by the relaxed design problem can be well explained as in above in the sense of engineering. Frankly speaking, the authors *did not expect* that the relaxed problem based on the homogenization method can provide this much detailed design. This is an example of power of the homogenization method.

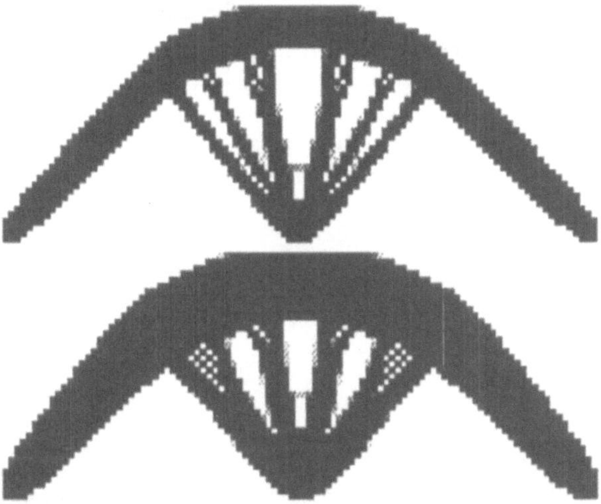

Figure 12 Optimal Designs of a Framed Structure

9. Structural Design for Multi-Loadings

In practice, a structure must often be designed under the multi-loading condition. That is, a structure must be safe for a set of various loadings. The present approach is easily extended to consider the multi-loading problem. Indeed, (1) with (6) is modified by

$$\begin{array}{l} \underset{d}{Minimize} \\ \text{subject to} \\ u^m \in V^m : a(u,v)=L^m(v)\ \forall v \in V_0^m \\ g_d(d) \le g_{d\max} \end{array} \quad \underset{m=1,\dots,M}{Maximize}\left\{ L^m(u^m) + \int_{\Gamma_d} \sum_{i,j=1}^{2} n_j \sigma_{ji}(u^m) u_{gi}^m d\Gamma \right\} \tag{12}$$

where

$$L^m(v) = \int_\Omega v^T b^m d\Omega + \int_{\Gamma_t} v^T t^m d\Gamma \tag{13}$$

and $V^m = \{\ v=\{v_1,v_2\} : v_i \in H^1(\Omega),\ v_i = u_{gi}^m \text{ on } \Gamma_g\ \}$, to consider M number of different loading cases $\{\ b^m, t^m, u_g^m\ \}$, $m=1,\dots,M$.

As an example of the multi-loading problem, we shall solve the design problem shown in Figure 13 in which a structure pin-supported at two different size circular holes is subject to three loadings, tension P1, bending downward P2, and bending upward P3. Furthermore, we shall restrict design to the case that parts of the constructed structure cannot pass through the center portion of the design domain.

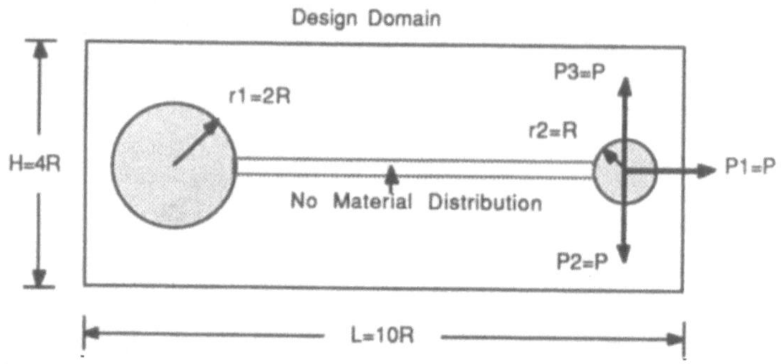

Figure 13. Design Domain, Restriction on Design, and Loadings

The relaxed design problem for the multi-loading problem (12) is solved using the similar homogenization method for (11), and the optimal designs are obtained as shown in Figure 14(a) for the case that we have the design restriction that no solid material is placed along the horizontal center line. Here two different volumes of the solid material are assumed. In Figure 14(b) the optimal designs are given for the case that such a design restriction is ignored.

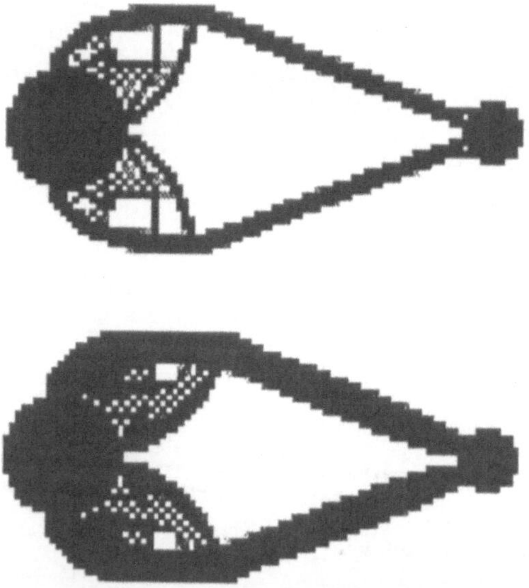

Figure 14(a) Optimal Designs for the Multi-Loading Problem
(with the Design Restriction)

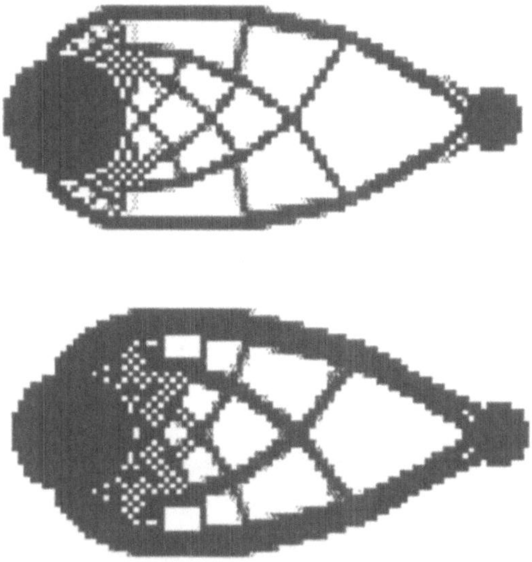

Figure 14(b) Optimal Design of the Multi-Loading Problem
(without the Design Restriction)

10. Extension to a Shell-like Structure

Extension of the relaxed design problem to other structural problems is also straight forward. As an example, we consider design of shell structures. In this case, the generalized state equation "$a(\boldsymbol{u},\boldsymbol{v}) = L(\boldsymbol{v})$" must be appropriately modified for the shell problem. Other than this modification, the same concept of the relaxed design problem is applicable.

As an example, a shell structure shown in Figure 15 is considered. This shell is supported at three points and is subject to a uniformly distributed transverse load. The design problem is that to find the optimal reinfocement by adding material on the already buildup thin shell structure, say 1mm thick shell. In some sense, this is same to the design problem to find the optimal thickness distribution of a shell. As shown in Figure 16, the optimal reinforcement is very discrete, i.e., there are little portion where perforated composite is assigned to minimize the mean compliance. That is, very clear rib reinfocement is obtained as the optimal solution. Basically two main ribs are generated along the lines connecting the three support points, while two rather narrow ribs are formed along the flat boundary edges to increse rigidity. If the shell is curved, the applied force is decomposed into the membrane and bending ones. That is, the curved edge can resist to transverse loads more than the flat edge. Thus reinforcement is requred along the flat edges of the shell considered. As shown in Figure 16, we have two ribs along these lines.

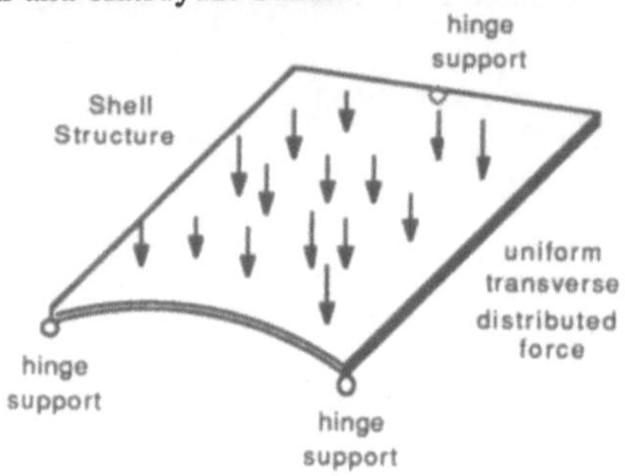

Figure 15 Design Domain of a Thin Shell Struatre

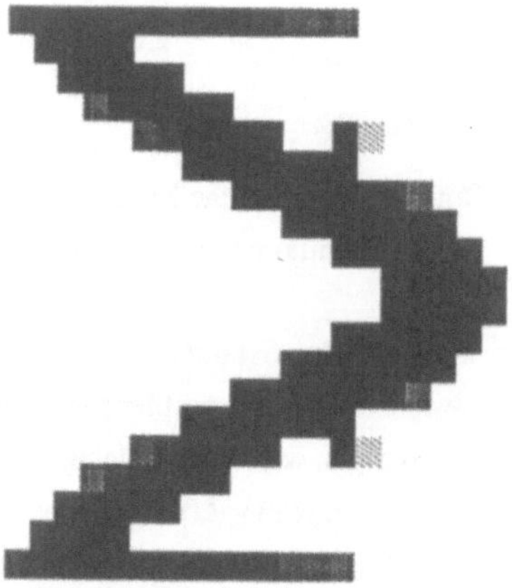

Figure 16 Top View of the Rib Reinforcement

11. Concluding Remarks

As shown in above, the relaxed design problem using the homogenization method can solve wide range design problems which have not been solved in engineering. This approach need not define the shape, topology, and size of a structure to determine their optimums, and is very different from usual engineering approaches available at present. Most of the difficulties we can find in usual engineering approaches cannot be found in the present method. The application of the homogenization method presented here to the structural optimization is merely an example to show the power of mathematics which has been developed and studied under the name of homogenization theory.

Acknowledgement

During the present work, the authors were supported by ONR N-00014-88-K-0637, DHHS-PHS-G-2-R01-AR34399-04, and RTB Corporation.

References

[1] Banichuk, N.V., *Problems and methods of optimal structural design*, Plenum Press, New York (1983)

[2] Cheng, K.T., and Olhoff, N., An investigation concerning optimal design of solid elastic plates, *Int. J. Solids and Structures* 17 (1981) 305-323

[3] Lurie, K.A., Fedorov, A.V., and Cherkaev, A.V., Regularization of optimal design problems for bars and plates, Parts I and II, *J. Optim. Theory Appl.* 37(4) (1982) 4999-521, 523-543

[4] Bendsøe, M.P., Generalized plate models and optimal design, in J.L. Eriksen, D. Kinderlehrer, R. Kohn and J.L. Lions, eds., *Homogenization and effect moduli of materials and media,* The IMA Volumes in Mathematics and Its Applications, Spriger-Verlag, Berlin, 1986, 1-26

[5] Palmer, A.C., Dynamic Programing and Structural Optimization, in R.H. Gallagher and O.C. Zienkiewicz, eds., *Optimum Structural Design*, John Wiley & Sons, Chichester (1973) 179-200

[6] Murat, F., and Tartar, L., Optimality conditions and homogenization, in A. Marino, L. Modica, S. Spagnolo, and M. Degiovanni, eds., *Nonlinear variational problems*, Pitman Advanced Publishing Program, Boston, 1985, 1-8

[7] Kohn, R., and Strang, G., Optimal Design and relaxation of variational problems, Parts I, II, and III, *Communications on Pure and Applied Mathematics*, XXXIX (1986) 113-137, 139-182, 353-378

[8] Bendsøe, M.P., and Kikuchi, N., Generating optimal topologies in structural design using a homogenization method, *Comput. Mechs. Appl. Mech. Engrg.*, 71 (1988) 197-224

[9] Suzuki, K., and Kikuchi, N., Shape and topology optimization using the homogenization method, (in review)*Comput. Mechs. Appl. Mech. Engrg.* (1989)

Noboru Kikuchi
Katsuyuki Suzuki
Department of Mechanical Engineering and Applied Mechanics
The University of Michigan
Ann Arbor, MI 48109
U.S.A.

Geometry and asymptotics in homogenization

S.M. KOZLOV

Abstract

We will discuss the methods for asymptotic computation of the effective parameters of a heterogeneous medium. Mostly, the case of strongly inhomogeneous media will be considered. The mathematical description of such media includes also a parameter which is responsible for the difference of the medium properties in distinct points. It will be demonstrated that inserting of this parameter shows clearly how the effective parameter depends on the structural medium geometry. This review relies on the work [1] and some further investigations.

General foundations

From the general point of view all the structures, which will be discussed below, are particular cases of random media. Homogenization for random media was worked out by the author in 1978 ([2]) and here it is necessary to recall the general result of [2].

Suppose that $(a_{ij}(x))_{i,j=1}^{n}$ is a random matrix valued function which is symmetric, ergodic and stationary in $x \in R^n$. Let A_ε be an operator with rapidly oscillating coefficients of the form

$$A_\varepsilon = \operatorname{div}(a(\frac{x}{\varepsilon})\operatorname{grad}) \ , \ a(x) = (a_{ij}(x)) \, . \qquad (0.1)$$

Then the basic result of [2] states that under an ellipticity condition on a , homogenization theory holds true for A_ε almost surely. The homogenized operator will be $\hat{A} = \operatorname{div}(\hat{a}\,\operatorname{grad})$, $\hat{a} = (a_{ij}) = \text{const}$, and can be found from the relation

$$\hat{a}(\xi) = \hat{a}_{ij}\xi_i\xi_j = \inf \langle a_{ij}(\xi_i + \frac{\partial}{\partial x_i}u)(\xi_j + \frac{\partial}{\partial x_j}u)\rangle . \qquad (0.2)$$

In (0.2) $\langle\cdot\rangle$ denotes the spatial mean of a random field, and the infimum is taken over all random stationary functions u . Actually, the infimum in (0.2) is not attained on some stationary field u . But there exists a stationary vector field $(v_1, \dots, v_n)$ such that

$$\begin{aligned} &\operatorname{div}(a(\xi + v)) = 0 , \\ &\frac{\partial}{\partial x_i} v_j = \frac{\partial}{\partial x_j} v_i \\ &\langle v\rangle = 0 . \end{aligned} \qquad (0.3)$$

Hence, in order to calculate $\hat{a}$ it is necessary to find a solution of (0.3) or some approximation of that solution. In one or other form this is done for all the media discussed below. But, instead of (0.2), in a number of examples it is better to use one of the following representations:

$$\hat{a}\xi = \langle a(\xi + v)\rangle \qquad (0.4)$$

or

$$\hat{a}\xi\,[\gamma] = \int_\gamma {*a(d(\xi\cdot x) + du)} \qquad (0.5)$$

(see [1], [12], [16], [17]). In (0.5) d is the differential, $*(dx_i) = (-1)^{i-1} dx_1 \wedge \overset{i}{\dots\wedge} \dots \wedge dx_n$, and γ is a hypersurface with homology class $[\gamma]$. Relation (0.5) is rigorous in the periodic case, when the (n–1) dimensional real homology of the torus is exactly R^n . For random structures (0.5) can't be used directly, but still it is an important intuitive recipe which is mostly useful in calculations, because the choice of the hypersurface γ is free.

1. Dispersed media

These are the inhomogeneous media with the simplest geometrical structure. The term dispersed medium means that inhomogeneity of such a structure is concentrated on sets which are isolated from each other, and are contained in some host medium with unconnected insertions. The first known example of such a medium was discussed by J.C. Maxwell, and then by numerous investigators beginning from J.W. Rayleigh. For a cubic periodic array of balls $B_0 = \{x : |x| < R\}$ of radius R, volume V, and for $a(x)$ in (0.1) given by

$$a(x) = \begin{cases} 0 \ , \exists m \in Z^3 \ , x \in B_0 + m \ , \\ 1 \ , \text{otherwise} \ , \end{cases} \tag{1.1}$$

an asymptotic expansion with respect to R of the homogenized operator is known, and in particular

$$\hat{a}_{ij} = \hat{a}\delta_{ij} \ , \qquad \hat{a} = \frac{1 - V}{1 + \frac{V}{2}} + O(R^{13}) \ , \tag{1.2}$$

where $R \to 0$ (see [3]). In [1] this formula was generalized for periodic systems of arbitrary bodies $B = RB_1$. If we pose $B_m = m + B$, $m \in Z^n$ (integer lattice in R^n) and define $a(x)$ as in (1.1), then instead of (1.2) we will get for the homogenized matrix $\hat{b}$ of such a structure

$$\hat{b} = (I - \frac{n-1}{n}(V + M))\,(I + \frac{1}{n}(V + M))^{-1} + O(R^{2n+2}) \ . \tag{1.3}$$

In (1.3) V is the volume of B, I is the identity matrix, and $M = (M_{ij})$ is the *added-mass* tensor of B (see [4]). The added-mass tensor is an important object in hydrodynamics and for many bodies it may be found explicitly. Note that (1.3) is nontrivial even in the case when $V = 0$. See, for example, Fig. 1.

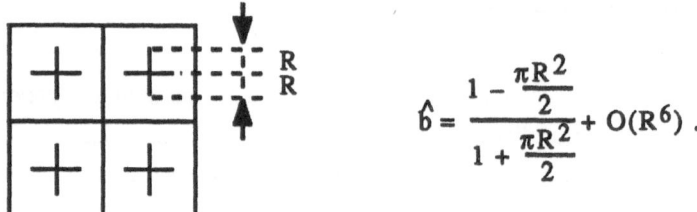

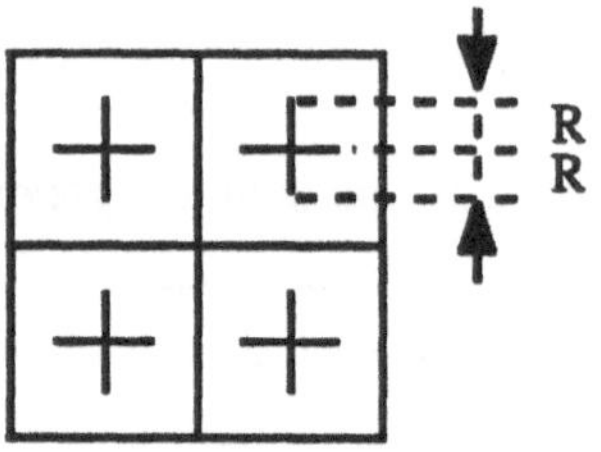

$$\hat{b} = \frac{1 - \frac{\pi R^2}{2}}{1 + \frac{\pi R^2}{2}} + O(R^6) \, .$$

FIGURE 1

In connection with (1.3) it is natural to remember Polya-Shiffer inequality ([3]). It says that among all the bodies with fixed volume the trace of the added-mass tensor is minimum for a ball. Then for $\hat{a}$ and $\hat{b}$ defined by (1.2) and (1.3) we obtain the following corollary as a consequence of (1.3) and of the cited inequality.

Corollary 1. *For any body* B_1 *there exists* $R_0 > 0$ *such that for* $R < R_0$, $B = RB_1$ *in (1.3) the following inequality holds:*

$$\frac{1}{3} \operatorname{tr} \hat{b} \leq \hat{a} \, . \tag{1.4}$$

What happens to $\hat{a}$ in (1.2), if the periodic array of balls is randomly perturbed? Suppose, for example, that the radii of the periodic system of balls have the form $R_m = r\xi_m$, where ξ_m, $m \in Z^3$, are independent identically distributed random variables. Denote the mean volume by $<V>$ and the variance of R_m^3 by $D(R^3) = <(R_m^3 - <R_m^3>)^2>$.

Theorem 1. *Let* $\hat{a}_r$ *be the homogenized coefficient of the random structure described above. Then*

$$\hat{a}_r = \frac{1 - \alpha <V>}{1 + \frac{\alpha}{2} <V>} + O(r^{11}) \tag{1.5}$$

$$\alpha = 1 + \zeta_6 D(R^3) \, ,$$

where ζ_6 *is the zeta function of the lattice*

$$\zeta_6 = \sum_{n \in Z^3} |n|^{-6} .$$

This and more general random balls systems are considered in [1]. Comparing (1.2) and (1.5) we see that in this model the following principle holds:

(Pr) *Randomization of the system decreases the effective parameter.*

To complete the section let us discuss the following question. Suppose that we compare configurations of balls with the same volume concentration. What configuration has the best effective parameter? To be precise suppose that in an arbitrary parallelogram $\Pi \subset R^2$ an arbitrary configuration of M distinct points $y = (y_1, \ldots, y_M)$ is given (see Fig. 2).

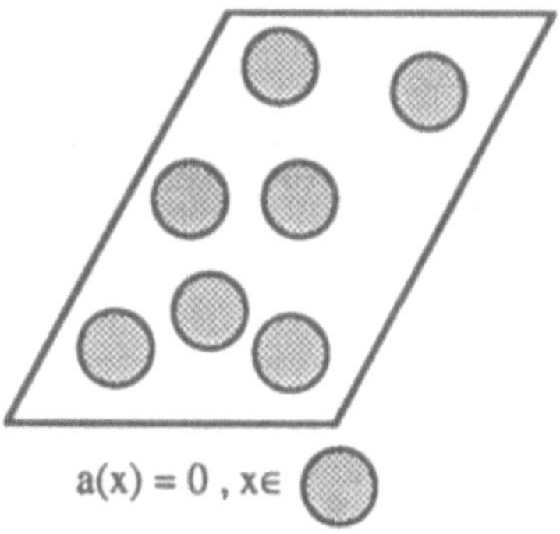

FIGURE 2
The parallelogram Π

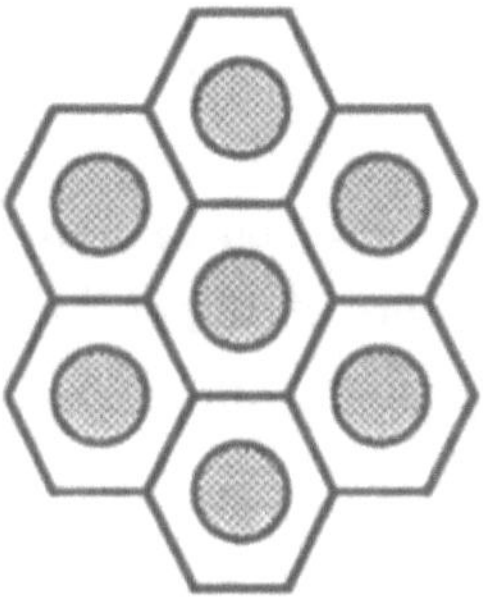

FIGURE 3
The hexagonal structure in two dimensions is the "champion"

Let, for $x \in \Pi$,

$$a(x) = \begin{cases} 0 , & \exists k = 1 , \ldots, M , |x - y_k| < R , \\ 1 & \text{otherwise} , \end{cases}$$

and set $\hat{\lambda}(y,R) = (\hat{a}_{11} + \hat{a}_{22})/2$, where $(\hat{a}_{ij})$ is the homogenized matrix for $a_{ij}(x) = a(x)\delta_{ij}$ given by (0.2) with periodic boundary conditions in Π . Let us also assume that λ_h is the effective parameter for the analogous hexagonal structure with the same volume fraction $(\frac{M\pi R^2}{|\Pi|})$ (see Fig. 3).

Theorem 2 ([1]). *For any* Π *and for any configuration of an arbitrary number* M *of balls, there exists* $R_0 > 0$ *such that for* $R < R_0$

$$\hat{\lambda}(y,R) \leq \lambda_h .$$

Here it is natural to remind that, by the theorem of A. Thue and L. Fejes Toth ([5]), the hexagonal packing is the closest packing of circles in the plane. This also means that the centers of the circles in the hexagonal packing are extremely far (in the mean) from each other in comparison with other configurations with the same concentration. So the statement of the theorem is in good accordance with geometrical intuition.

Theorem 2 is another realization of (Pr). In general, the problem that extremality of effective characteristics entails order (periodicity) in the medium seems to be a deep one. For example, in the proof of the Theorem 2 fine facts from the theory of elliptic and modular function are used. The problem has also a connection with the earlier geometrical theories of A. Schoenflies and B.N. Delone [6].

2. Chess structures

Suppose that a chess structure is given as shown in Fig. 4 with

$$a(x) = \begin{cases} 1 & \text{on white cells,} \\ \delta & \text{on black cells.} \end{cases}$$

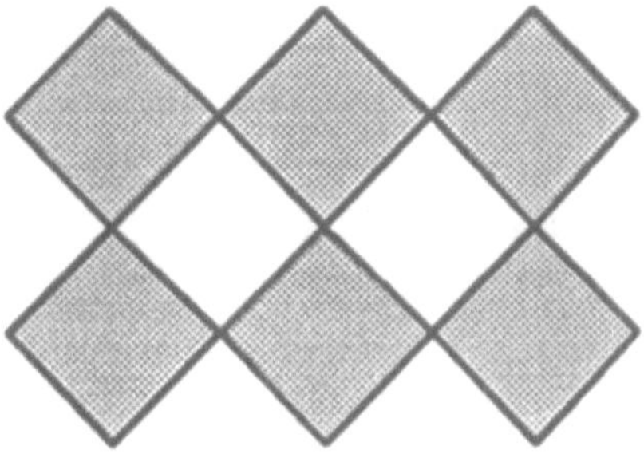

$\hat{a} = \sqrt{1 \cdot \delta}$

FIGURE 4

Then the exact answer obtained in [7] says that $\hat{a} = \sqrt{1 \cdot \delta}$. Basing on this example, in [8] a general algebraic construction was established based on the Legendre transform on cohomologies. This construction has led to a set of extra explicit formulas ([8], [3]).

But it seems that explicit formulas are available only as exceptions. In order to clarify the situation it is natural to consider δ as a small parameter. Then, for example, for rhombus structures

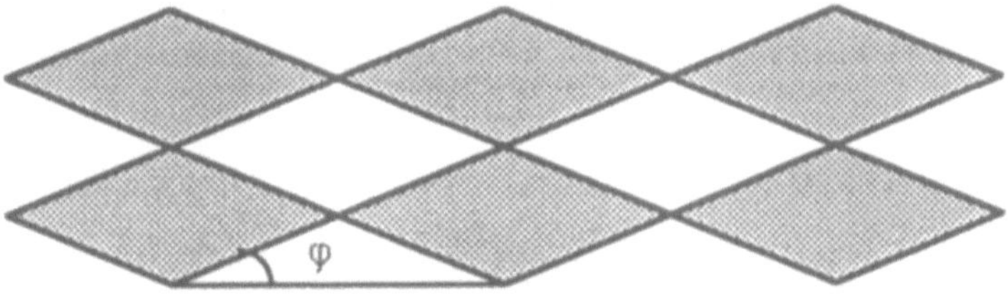

FIGURE 5

we have

$$\hat{a}_1 = \operatorname{ctg}\varphi \sqrt{\frac{\varphi}{\pi/2 - \varphi}} \sqrt{\delta}\,(1 + O(\sqrt{\delta}))\,,$$

$$\hat{a}_2 = \operatorname{tg}\varphi \sqrt{\frac{\pi/2 - \varphi}{\varphi}} \sqrt{\delta}\,(1 + O(\sqrt{\delta}))\,.$$

Now suppose that the chess structure shown in Fig. 4 is an elastic medium, where white cells have Lamé coefficients λ, μ and the black cells have Lamé coefficients $\delta\lambda, \delta\mu$ and both are homogeneous and isotropic (the case was worked out jointly with L.V. Berlyand (see[9])). Thanks to structure symmetry the homogenized elastic energy density has the form

$$J(e) = \frac{1}{2}\hat{a}_1(e_{11}^2 + e_{22}^2) + \hat{a}_2 e_{11} e_{22} + 2\hat{a}_3 e_{12}^2 , \tag{2.1}$$

$$e_{ij} = \frac{1}{2}\left(\frac{\partial}{\partial x_i} u_j + \frac{\partial}{\partial x_j} u_i\right) , \quad i, j = 1,2 ,$$

so there are three effective parameters $\hat{a}_i$. The homogenized medium can also be represented with the three following parameters: Young and shear moduli $\hat{E}$ and $\hat{\mu}$, and Poisson's ratio $\hat{\nu}$. Then

$$\hat{E} = e\sqrt{\delta}\,(1 + O(\sqrt{\delta})) , \quad \hat{\mu} = m\sqrt{\delta}\,(1 + O(\sqrt{\delta})) ,$$
$$\hat{\nu} = O(\delta^{1/4-\varepsilon}) , \quad \varepsilon > 0 , \tag{2.2}$$

where the constants e and m can be found explicitly (see [9]). From the first sight it's amazing that Poisson's ration is vanishing with δ . (Note that Poisson's ratio of white and black cells are equal and independent of δ !).

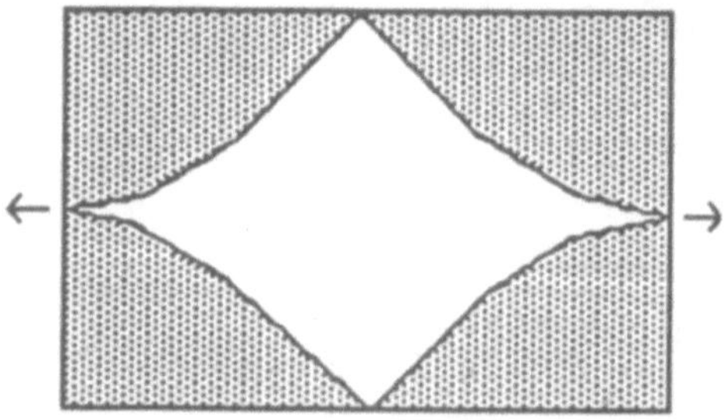

Young modulus $\hat{E}$
Poisson's ratio $\hat{\nu}$

FIGURE 6

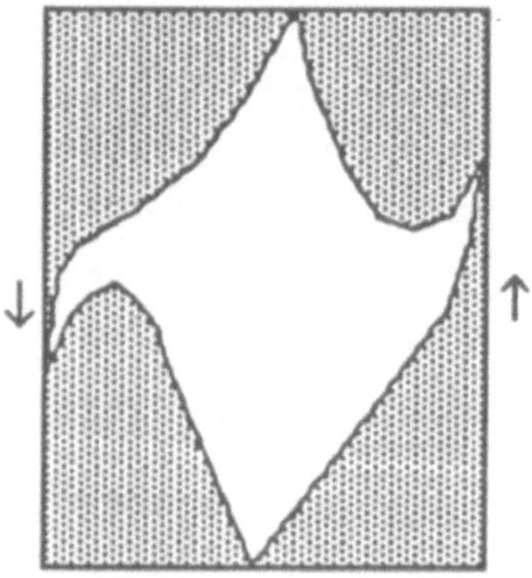

Shear modulus $\hat{\mu}$

FIGURE 7

But looking at Fig. 6 we see that, when the load is horizontal, only one kind of angles (the "noses") really works. This means that pratically all the deformation is made thanks to "noses" and compression in the other direction is negligible.

From Fig. 6 and 7 it is rather clear that effective coefficients are asymptotically determined by the singularities of the solution of the elastic analogue of (0.3) in the corresponding angle point. The analysis of these singularities is made in [9].

Coming back to the scalar case, let's consider three-dimensional cubic structures as in Fig. 8 and 9.

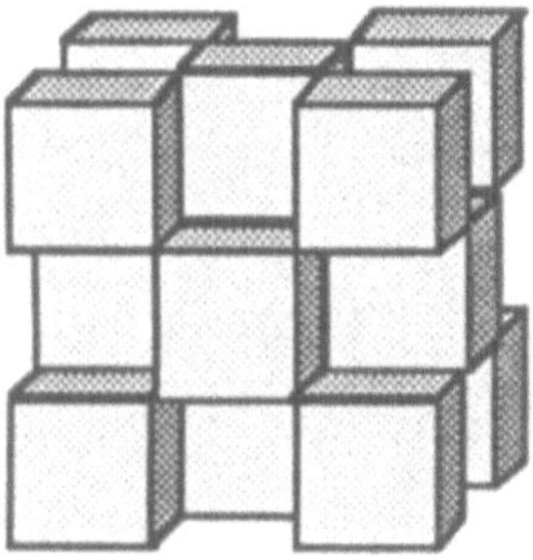

FIGURE 8

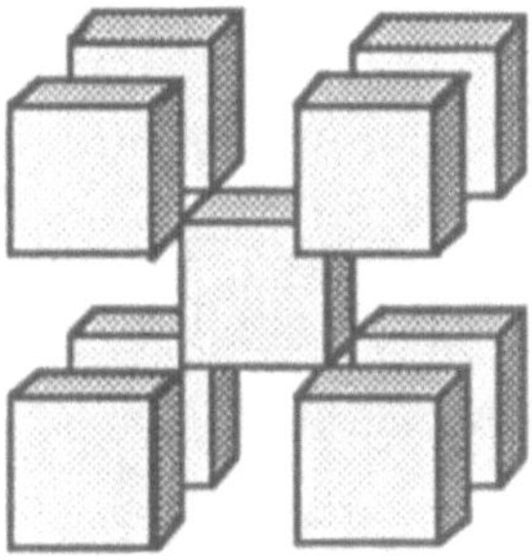

FIGURE 9

It is convenient to speak about conductivity of such structures. Suppose that

$$a(x) = \begin{cases} 1 & \text{on hatched cubes,} \\ \delta & \text{otherwise .} \end{cases} \qquad (2.3)$$

Due to the symmetry, the effective conductivity in both cases is scalar, and it can be shown that

$$\hat{a} \sim c_1 \sqrt{\delta} \qquad \text{(Fig. 8)} \qquad (2.4)$$

$$\hat{a} \sim c_2\, \delta . \qquad \text{(Fig. 9)} \qquad (2.5)$$

This result of [1] means that there are no other laws for three dimensional linear conductivity. Laws of different type can be obtained for nonlinear conductivity in any dimension $n > 2$. Let us discuss the homogenization problem:

$$\hat{a}_\delta(\xi) = \inf_u \int_{T^n} a(x)\, |\xi + \nabla u|^\alpha dx\,, \qquad \alpha > 1\,, \tag{2.6}$$

where $a(x)$ is taken from Fig. 9 and (2.3). Then, for $1 < \alpha < n$, $\hat{a}_\delta(\xi) \sim c(\xi)\cdot\delta$; and $\hat{a}_\delta(\xi) \sim c(\xi) > 0$, $n < \alpha < \infty$. If $\alpha = n$, then $\hat{a}_\delta(\xi)$ can be estimated as follows for $|\xi| = 1$

$$c_2\,\delta^{\frac{n-1}{n}} \le \hat{a}_\delta(\xi) \le c_1\,\delta^{\frac{n-1}{n}}\,, \tag{2.7}$$

$0 < c_2 < c_1 < \infty$. Inequalities (2.7) are an asymptotic generalization of the geometrical mean law to arbitrary dimension n. The opposite case of small fluctuations of $a(x)$ for (2.6) (i.e. $\delta \sim 1$) was considered in [10].

3. Disordered chess structures

These structures and networks similar to them were a subject of numerous physical and computational investigations (see [11] and literature quoted therein). Here we present some rigorous results from [1]. For analogous results for networks of random resistors see [12]. Let us consider a plane R^2 covered with the regular D-polygons, D = 3, 4, 6 (see Fig. 10).

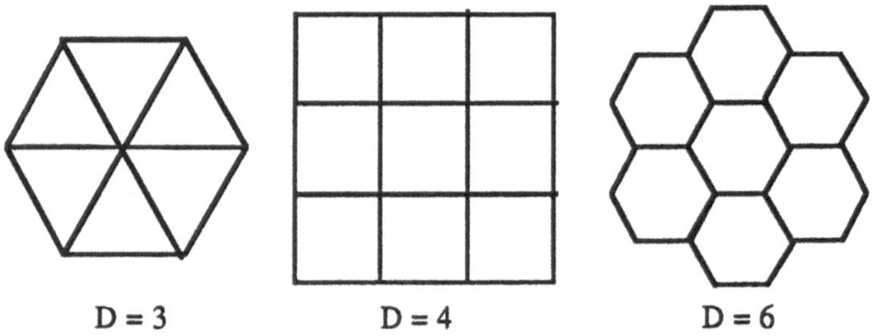

FIGURE 10

Suppose that the colors of the cells are random and independent: black with probability p and white with probability $1-p$, $0 \le p \le 1$. Again, let's introduce the random function:

$$a(x) = \begin{cases} 1 & \text{on white cells} \\ \delta & \text{on black cells} \end{cases} \tag{3.1}$$

and denote by $a(p,\delta)$ the respective homogenized coefficient, $0 \le \delta \le 1$. For $a(p,\delta)$ there is a relation (see [2])

$$a(p,\delta)\, a(1-p,\delta) = \delta . \tag{3.2}$$

We are concerned in the properties of $a(p,\delta)$, beginning from small p.

Theorem 3 ([1]). *The function* $a(p,\delta)$ *has a derivative with respect to* p *at* $p = 0$. *Moreover*

$$a(p,\delta) = 1 - \alpha(\delta)\, p + O(p^2) \tag{3.3}$$

uniformly with respect to $\delta \ge 0$ *and*

$$-\frac{\partial}{\partial p} a(p,0) = \alpha(0) = \frac{\tan(\pi D^{-1})}{2\pi D}\, \frac{\Gamma^4(D^{-1})}{\Gamma^2(2D^{-1})} , \tag{3.4}$$

where $\Gamma(\cdot)$ *is Euler's gamma-function.*

Now we will discuss the behaviour of $a(p,\delta)$ for $\delta \to 0$ and $0 \le p \le 1$. To do that, let us point out the centers of the cells and connect the nearest neighbours. In this way we obtain a network on R^2. We will denote the percolation probability of the site problem on this network by $1-p_D$ (see [13]). Remind that

$$p_3 = 2\sin\frac{\pi}{18} , \quad p_4 \approx 0.41 , \quad p_6 = 0.5 . \tag{3.5}$$

Theorem 4 ([1]). *For* $0 \leq \delta \leq 1$, $a(p,\delta)$ *is a monotone function of* p *and for suitable constants* $0 < c_i < \infty$, $i = 1, \ldots, 4$, *we have*

$$\begin{aligned} &1. \quad c_1 < a(p,\delta) < 1 && 0 < p < p_D , \\ &2. \quad c_2\sqrt{\delta} < a(p,\delta) < c_3\sqrt{\delta} && p_D < p < 1 - p_D , \\ &3. \quad \delta < a(p,\delta) < c_4\,\delta && 1 - p_D < p < 1 . \end{aligned} \tag{3.6}$$

As a consequence of (3.6) we have logarithmic asymptotics

$$a(p,\delta) \sim \begin{cases} 1 & 0 \leq p < p_D \\ \sqrt{\delta} & p_D < p < 1 - p_D \\ \delta & 1 - p_D < p \leq 1 . \end{cases} \tag{3.7}$$

This means that for $D = 3, 4$ the exact value $a(1/2,\delta) = \sqrt{\delta}$ is valid asymptotically in the whole band of probabilities $p_D < p < 1 - p_D$.

For Theorems 3, 4 basic results from rigorous percolation theory are essential. To explain the ideas of the proof of Theorem 4 for $D = 4$ it is necessary to introduce two types of connections between cells:

(I) cells are in contact if they have a common edge,

(II) cells are in contact if they have a common point .

Then, according to percolation theory, for $p < p_4$ there exist infinite chains of white cells connected in a first way, and this geometric fact leads to (3.6.1). When $p > p_4$ there are no first type white chains nor black (until $p < 1 - p_4$). But in the band $(p_4, 1 - p_4)$ there coexist infinite white and black chains of the second type. Look at Fig. 11, where $N \to \infty$.

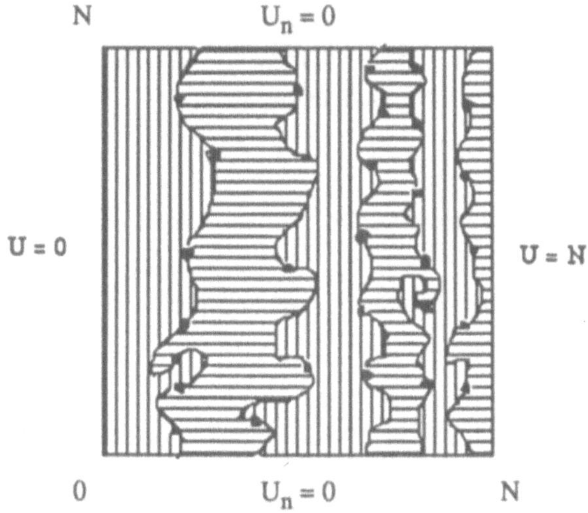

FIGURE 11

"Vertical" black chains of second type divide the square into bars. In each bar the potential U (solution of the problem (0.2)) practically is constant. In each chain there are a number of "hot" points (pointed out on Fig. 11) in which the black cells *contact only with the angles* . These points are the points of overdistribution of the potential ($|\nabla U|$ has singularity in them so they are "hot"). Potential U near "hot" points can be found approximately as follows:

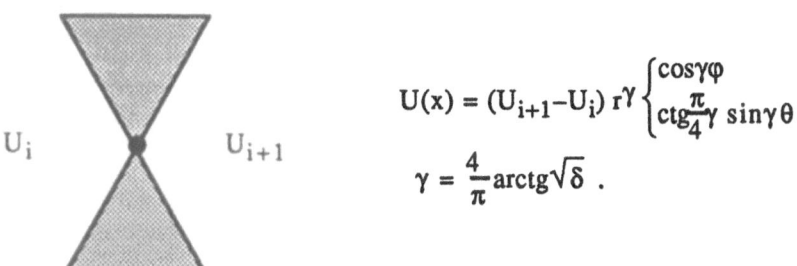

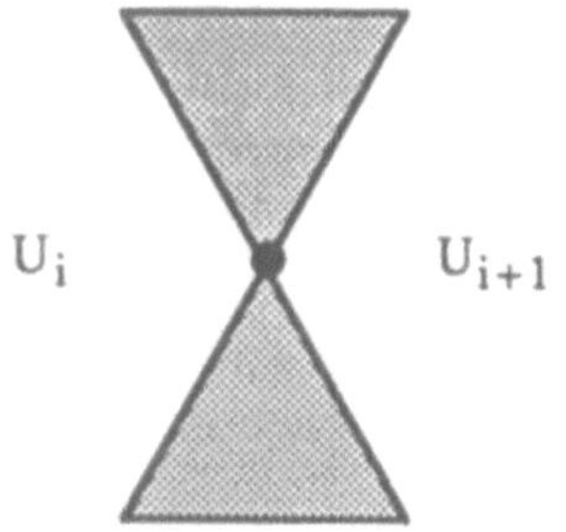

FIGURE 12

$$U(x) = (U_{i+1} - U_i)\, r^{\gamma} \begin{cases} \cos\gamma\varphi \\ \operatorname{ctg}\frac{\pi}{4}\gamma \,\sin\gamma\theta \end{cases}$$

$$\gamma = \frac{4}{\pi}\operatorname{arctg}\sqrt{\delta} \, .$$

Here U_i, U_{i+1} are constant values of the potential on the "left" and "right" sides of the chain. When p passes through $1 - p_4$, black chains of first type begin to exist and hence $a(p,\delta) \sim \delta$ in this band.

4. Strongly inhomogeneous continuous media

Here a topology free extract of [1], [14] will be given on some examples. Let us consider the following operator

$$A = \mathrm{div}(e^{\lambda S(x)}\,\mathrm{grad})\,. \tag{4.1}$$

In (4.1) $S(x) \in C^\infty$ is a Z^n-periodic function with nondegenerate critical points and distinct critical values (a so called Morse function). Denote by $(\hat{a}_{ij}(\lambda))$ the matrix of homogenized coefficients for A, $\hat{a}_\lambda(\xi) = \hat{a}_{ij}\xi_i\xi_j$. We are interested in the asymptotics of $\hat{a}_\lambda(\xi)$ when $\lambda \to \infty$. To formulate the result it is necessary to do some preliminary geometrical constructions.

Let us introduce the *first percolation level* S_1 :

> S_1 *is the greatest real number* h *such that the set* $\{x \in R^n : S(x) \geq h\}$ *contains an unbounded connected component*.

Let us also introduce the *first percolation cycle* m :

> *first nonzero integer vector* $m = (m_1, \ldots, m_n)$ *such that* $\{x : S(x) = S_1\}$ *can be connected with* $x + m$ *by a path* $\gamma(t)$ *such that* $S(x)\big|_{\gamma(t)} \geq S_1$.

From the Morse theory it follows that on the level S_1 there is a unique critical point of $S(x)$, say x_1 and in some orthonormal coordinate system $\{y_i\}$ with the origin at the point x_1

$$S(y) = S_1 + 1/2(K_1 y_1^2 - K_2 y_2^2 - \ldots K_n y_n^2) + O(|y|^3)\,,$$

with $K_i > 0$, $i = 1, \ldots, n$.

Theorem 5. *For* $\hat{a}(\xi)$ *the following asymptotic expansion holds:*

$$\hat{a}(\xi) \sim \left(\frac{2\pi}{\lambda}\right)^{\frac{n-2}{2}} \sqrt{\frac{K_1}{K_2 \cdot \ldots \cdot K_n}}\,(m\cdot\xi)^2\, e^{\lambda S_1}\left(1 + \sum_{j=1}^{\infty} c_j \lambda^{-j}\right) \tag{4.2}$$

where $m = (m_1, \ldots, m_n)$ *is the first percolation cycle and* $m\cdot\xi$ *is the scalar product of* m *and* ξ.

This theorem and the related ideas can be used successfully in homogenization of diffusion problems and for the so called effective diffusion (see [14]-[17]).

The first term of formula (4.2) was obtained in [1] in the context of *variational Laplace integrals*. This concept includes also a great deal of nonlinear homogenization problems. Let us discuss the following nonlinear problem

$$\hat{f}_\lambda(\xi) = \inf_{u \in C^\infty(T^n)} \int_{T^n} e^{\lambda S(x)} f(x, |\xi + \nabla u|)\,dx\,, \tag{4.3}$$

where $|\cdot|$ is the Euclidean norm, $T^n = R^n/Z^n$, $f(x,t)$ is convex in t and satisfies some technical conditions omitted here (see [1]). Denote by f^* the Legendre transform of f and

$$\bar{f}(t) = (2\pi)^{-1/2} \int_{-\infty}^{+\infty} e^{z^2/2} f(te^{-z^2/2})\,dz\,.$$

Theorem 6. *Let* $S(x)$ *be a Morse function and let* m, *and* S_1 *be as above. Then*

$$f(\xi) = \frac{1}{\sqrt{K_1 \cdot \ldots \cdot K_n}} \left(\frac{2\pi}{\lambda}\right)^{\frac{n}{2}} \cdot \bar{f}^{**}\left(\sqrt{\frac{\lambda K_1}{2\pi}}\; m\cdot\xi\right) e^{\lambda S_1}(1 + o(1))\,. \tag{4.4}$$

This theorem has the following application to the homogenization of a hydrodynamical filtration problem posed in [8]. Suppose that $p(x) \in C^\infty(T^n)$,

$p(x) > 1$ and that at the first percolation level $\bar{p}$ there are n independent percolation cycles $m^1, \ldots, m^n \in Z^n$ (that means *percolation over all directions* at the same level). Let

$$\hat{f}(\xi) = \inf_{u \in C^\infty(T^n)} \int_{T^n} |\nabla u + \xi|^{p(x)} dx .$$

Then from Theorem 6 we obtain, as $|\xi| \to \infty$,

$$\hat{f}(\xi) \sim c_1(\omega) \, |\xi|^{\bar{p}} (\ln|\xi|)^{\frac{\bar{p}-n}{2}} ,$$

where $\omega = \xi / |\xi|$.

Acknowledgments. I wish to thank Professor Gian Fausto Dell'Antonio for his hospitality during my stay at the International Centre for Theoretical Physics of Trieste, where this review was written. I am pleased to thank Miss Claudia Parma for her help in preparing the manuscript.

REFERENCES

[1] S.M. Kozlov, *Geometric aspects of homogenization*, Russian Math. Surveys **40** (1989), 79-120.

[2] S.M. Kozlov, *The averaging of random structures*, Docl. Akad. Nauk. SSSR, **241** (1978), 1016-1019, Matem. Sbornik, 1979, 199-212.

[3] V.L. Berdicevskii, *Variational principle in spatial medium*, M. "Nauka", 1983 (In Russian).

[4] J.M. Newman, *Marine Hydrodynamic*, MIT Press, 1977.

[5] C.A. Rogers, *Packing and Covering*, Cambridge Univ. Press, 1964.

[6] R.V. Galiulin, *Crystallographic geometry*, M. 1985 (In Russian).

[7] J.B. Keller , *A theorem on the conductivity of a composite medium*, J. Math. Phys. **5**, 548-549.

[8] S.M. Kozlov, *Duality of one type of variational problems*, Functional Anal. Appl. **17** (1983), 171-175.

[9] L.V. Berlyand, S.M. Kozlov, *Asymptotic of effective moduli for elastic chess composite* , (in publication).

[10] R. Blumenfeld, D.J. Bergman, *Exact calculation to second order of effective dielectric constant of a strongly nonlinear inhomogeneous composite*, Physical Review B, **40**, 1989.

[11] D.J. Bergman, *Elastic moduli near percolation in two dimensional random network of rigid and nonrigid bounds*, Physical Review B, **33**, 1986, 2013-2016.

[12] S.M. Kozlov, *Homogenization for disordered systems*, Thesis, Moscow, 1988.

[13] H. Kesten, *Percolation theory for mathematicians*, Birkhäuser Boston Inc., 1982.

[14] S.M. Kozlov, *Asymptotics of the Laplace-Dirichlet integrals*, Funct. Anal. Appl. **40**, 1990.

[15] S.M. Kozlov, *Reducability and averaging of quasiperiodic operators*, Transactions of Moscow Mathematical Society **46**, 1983.

[16] S.M. Kozlov, *Random walks and averaging in inhomogeneous media*, Russian Math. Survey **40**, 1985, 61-120.

[17] S.M. Kozlov, *Effective diffusion for Focker-Plank equation*, Matem. Zametki **45**, 1989, 19-31.

S.M. Kozlov
Moscow Institute of Civil Engineering
Moscow 129 337
Yaroslavskoe Chausse, 26

The Field Equation Recursion Method

GRAEME W. MILTON

Introduction

Effective tensors, such as the effective conductivity tensor σ^* and the effective elasticity tensor $\mathbf{C}^*$ govern the macroscopic response of composites to applied fields. These tensors are strongly influenced by the details of the microgeometry and considerable emphasis has been placed on deriving microstructure independent equalities or inequalities (bounds) on effective tensors. This paper briefly reviews some of the various methods for bounding effective tensors of two-component composites, and reviews the significance of an associated tensor $\mathbf{Y}^*$, obtained from σ^* or $\mathbf{C}^*$ via a fractional linear tensor transformation. We also draw attention to the recent work of Cherkaev and Gibiansky [1] who independently introduced the tensor $\mathbf{Y}^*$ as an aid in analyzing the bounds of the translation method. Their work adds substantial weight to the growing body of evidence which suggests that the tensor $\mathbf{Y}^*$ is fundamentally important. The field equation recursion method, discussed in Section 5, gives the tensor $\mathbf{Y}^*$ a physical interpretation and thus links together the various bounding methods.

To bound the effective conductivity tensor it suffices to consider periodic media with periodic fields. The equations of electrical conductivity

$$\mathbf{j} = \sigma\mathbf{e}, \ \nabla \times \mathbf{e} = 0 \ , \ \nabla \cdot \mathbf{j} = 0, \tag{1.1}$$

when solved for a basis set of driving fields

$$\mathbf{e}^* = ⨍\mathbf{e}(\mathbf{x})d\mathbf{x} \equiv \langle \mathbf{e} \rangle , \tag{1.2}$$

give the effective tensor σ^* via its defining relation

$$\mathbf{j}^* = \sigma^* \mathbf{e}^* \ \text{ where } \ \mathbf{j}^* = \langle \mathbf{j} \rangle . \tag{1.3}$$

Here the angular brackets denote volume averages over the unit cell of periodicity. For a two-component material with isotropic phases local conductivity takes the form

$$\sigma(\mathbf{x}) = \sigma_1\chi_1(\mathbf{x})\mathbf{I} + \sigma_2\chi_2(\mathbf{x})\mathbf{I}, \tag{1.4}$$

where

$$\chi_1(\mathbf{x}) = 1 - \chi_2(\mathbf{x}) = \begin{cases} 1 \text{ in component } 1 \\ 0 \text{ in component } 2 \end{cases} \tag{1.5}$$

is the characteristic function describing the microstructure of the composite. The averages of these characteristic functions

$$f_1 = <\chi_1>; \; f_2 = <\chi_2> = 1 - f_1 \tag{1.6}$$

are of course just the volume fraction of the components.

2. VARIATIONAL PRINCIPLES AND THE BERAN AND HASHIN-SHTRIKMAN BOUNDS

Estimates of the effective conductivity $\sigma^*\mathbf{I}$ of a two-component isotropic composite can be directly obtained from energy minimization principles. The standard approach is to use the Dirichlet and Thompson variational principles,

$$\mathbf{e}^*\cdot\sigma^*\mathbf{e}^* = \min_{\substack{\mathbf{e}(\mathbf{x}) \\ <\mathbf{e}> = \mathbf{e}^* \\ \nabla\times\mathbf{e}(\mathbf{x}) = 0}} <\mathbf{e}(\mathbf{x})\cdot\sigma(\mathbf{x})\mathbf{e}(\mathbf{x})>, \tag{2.1}$$

$$\mathbf{j}^*\cdot\sigma^{*-1}\mathbf{j}^* = \min_{\substack{\mathbf{j}(\mathbf{x}) \\ <\mathbf{j}> = \mathbf{j}^* \\ \nabla\cdot\mathbf{j}(\mathbf{x}) = 0}} <\mathbf{j}(\mathbf{x})\cdot\sigma(\mathbf{x})^{-1}\mathbf{j}(\mathbf{x})>. \tag{2.2}$$

By taking constant trial fields $\mathbf{e}(\mathbf{x}) = \mathbf{e}^*$ and $\mathbf{j}(\mathbf{x}) = \mathbf{j}^*$ the variational principles give the arithmetic and harmonic mean bounds

$$\sigma^* \le f_1\sigma_1 + f_2\sigma_2, \tag{2.3}$$

$$\sigma^{*-1} \le f_1/\sigma_1 + f_2/\sigma_2, \tag{2.4}$$

on the effective conductivity σ^*. To obtain better estimates from these variational principles we need to take trial fields which are correlated with the geometry of the composite, i.e. that depend on the characteristic function $\chi_1(\mathbf{x})$.

Following Beran [2] let us take a trial field of the form

$$\mathbf{e}(\mathbf{x}) = \mathbf{e}^* + \alpha \mathbf{e}_1(\mathbf{x}), \tag{2.5}$$

where $\mathbf{e}_1(\mathbf{x})$ is the field satisfying

$$<\mathbf{e}_1> = 0, \quad \nabla \times \mathbf{e}_1(\mathbf{x}) = 0\,, \quad \nabla \cdot \mathbf{e}_1(\mathbf{x}) = \nabla \cdot \chi_1(\mathbf{x})\mathbf{e}^*. \tag{2.6}$$

Such trial fields are appropriate because they match the true field in the composite when the material is nearly homogeneous, i.e. when $\sigma_1 - \sigma_2$ is small. Substituting the trial field in the Dirichlet variational principle gives

$$\mathbf{e}^* \cdot \sigma^* \mathbf{e}^* \le \min_{\alpha} \left[\mathbf{e}^* \cdot <\sigma> \mathbf{e}^* + 2\alpha <\mathbf{e}^* \cdot \sigma \mathbf{e}_1> + \alpha^2 <\mathbf{e}_1 \cdot \sigma \mathbf{e}_1> \right]$$

$$\le (f_1\sigma_1 + f_2\sigma_2)\mathbf{e}^* \cdot \mathbf{e}^* - <\mathbf{e}^* \cdot \sigma \mathbf{e}_1>^2 / <\mathbf{e}_1 \cdot \sigma \mathbf{e}_1>. \tag{2.7}$$

To simplify the bound further we note that the defining equations (2.6) for $\mathbf{e}_1(\mathbf{x})$ imply its Fourier transform is simply

$$\hat{\mathbf{e}}_1(\mathbf{k}) = \frac{\mathbf{k}(\mathbf{k} \cdot \mathbf{e}^*)}{k^2} \hat{\chi}_1(\mathbf{k}) \quad \text{when } \mathbf{k} \ne 0$$

$$= 0 \quad \text{when } \mathbf{k} = 0, \tag{2.8}$$

where $\hat{\chi}_1(\mathbf{k})$ is the Fourier transform of $\chi_1(\mathbf{x})$. By Plancherel's formula it follows that

$$<\mathbf{e}^* \cdot \sigma \mathbf{e}_1> = (\sigma_1 - \sigma_2) <\mathbf{e}^* \chi_1 \mathbf{e}_1>$$

$$= (\sigma_1 - \sigma_2) \sum_{\mathbf{k} \ne 0} \frac{(\mathbf{k} \cdot \mathbf{e}^*)^2}{k^2} \hat{\chi}_1(\mathbf{k}) \hat{\chi}_1(-\mathbf{k}), \tag{2.9}$$

$$<\mathbf{e}_1 \cdot \sigma \mathbf{e}_1> = \sigma_2 <\mathbf{e}_1 \cdot \mathbf{e}_1> + (\sigma_1 - \sigma_2) <\mathbf{e}_1 \cdot \chi_1 \mathbf{e}_1>$$

$$= \sigma_2 \sum_{\mathbf{k} \ne 0} \frac{(\mathbf{k} \cdot \mathbf{e}^*)^2}{k^2} \hat{\chi}_1(\mathbf{k}) \hat{\chi}_1(-\mathbf{k})$$

$$+ (\sigma_1 - \sigma_2) \sum_{\mathbf{k} \ne 0} \sum_{\mathbf{m} \ne 0} \frac{(\mathbf{e}^* \cdot \mathbf{m})(\mathbf{m} \cdot \mathbf{k})(\mathbf{k} \cdot \mathbf{e}^*)}{m^2 k^2} \hat{\chi}_1(\mathbf{k}) \hat{\chi}_1(\mathbf{m} - \mathbf{k}) \hat{\chi}_1(-\mathbf{m}), \tag{2.10}$$

where the sums extend over all lattice points in Fourier space excluding the origin. Since the geometry of the composite is isotropic, we can average these sums over the direction of $\mathbf{e}^*$ to obtain

$$\sum_{\mathbf{k}\neq 0}\frac{(\mathbf{k}\cdot\mathbf{e}^*)^2}{k^2}\hat{\chi}_1(\mathbf{k})\hat{\chi}_1(-\mathbf{k})=\sum_{\mathbf{k}\neq 0}\hat{\chi}_1(\mathbf{k})\hat{\chi}_1(-\mathbf{k})\mathbf{e}^*\cdot\mathbf{e}^*/d$$

$$=<(\chi_1(\mathbf{x})-f_1)^2>\mathbf{e}^*\cdot\mathbf{e}^*/d$$

$$=f_1f_2\mathbf{e}^*\cdot\mathbf{e}^*/d, \tag{2.11}$$

$$\sum_{\mathbf{k}\neq 0}\sum_{\mathbf{m}\neq 0}\frac{(\mathbf{e}^*\cdot\mathbf{m})(\mathbf{m}\cdot\mathbf{k})(\mathbf{k}\cdot\mathbf{e}^*)}{m^2k^2}\hat{\chi}_1(\mathbf{k})\hat{\chi}_1(\mathbf{m}-\mathbf{k})\hat{\chi}_1(-\mathbf{m})=A\mathbf{e}^*\cdot\mathbf{e}^*/d, \tag{2.12}$$

where d is the dimension of space and

$$A=\sum_{\mathbf{k}\neq 0}\sum_{\mathbf{m}\neq 0}\frac{(\mathbf{m}\cdot\mathbf{k})^2}{m^2k^2}\hat{\chi}_1(\mathbf{x})\hat{\chi}_1(\mathbf{m}-\mathbf{k})\hat{\chi}_1(-\mathbf{m}) \tag{2.13}$$

is a parameter which depends only on the composite geometry. Substituting these expressions in the bound gives

$$\sigma^*\leq f_1\sigma_1+f_2\sigma_2-\frac{f_1f_2(\sigma_1-\sigma_2)^2}{f_2\sigma_1+f_1\sigma_2+(d-1)(\zeta_1\sigma_1+\zeta_2\sigma_2)}, \tag{2.14}$$

where the parameters

$$\zeta_1=(\frac{Ad}{f_1f_2}-f_2)/(d-1)\ ,\ \zeta_2=1-\zeta_1, \tag{2.15}$$

have been introduced because they satisfy simple inequalities and because they make the bound (2.14) look more pleasing. The positivity of

$$<(\mathbf{e}^*+\alpha\mathbf{e}_1)\cdot\chi_1(\mathbf{e}^*+\alpha\mathbf{e}_1)>=f_1+2f_1f_2\alpha/d+\alpha^2A/d, \tag{2.16}$$

for all α, and in particular for $\alpha=-f_1f_2/A$ which is the value of α that minimizes (2.16), implies

$$\zeta_1\geq 0\ , \tag{2.17}$$

and similarly the positivity of

$$<(\mathbf{e}^*+\alpha\mathbf{e}_1)\cdot\chi_2(\mathbf{e}^*+\alpha\mathbf{e}_1)>=f_2-2f_1f_2\alpha/d+\alpha^2(f_1f_2-A)/d, \tag{2.18}$$

implies

$$\zeta_1\leq 1. \tag{2.19}$$

The form (2.14) of the Beran bound (see also Milton [3] and Torquato [4]) suggests it is appropriate to introduce a variable Y^* defined by

$$\sigma^* = f_1\sigma_1 + f_2\sigma_2 - \frac{f_1 f_2(\sigma_1 - \sigma_2)^2}{f_2\sigma_1 + f_1\sigma_2 + Y^*}, \tag{2.20}$$

The upper bound (2.14) together with the associated lower bound (obtained in a similar manner but from the Thompson variational principle (2.2)) imply

$$(d-1)(\zeta_1/\sigma_1 + \zeta_2/\sigma_2)^{-1} \leq Y^* \leq (d-1)\,(\zeta_1\sigma_1 + \zeta_2\sigma_2). \tag{2.21}$$

Thus apart from a normalization factor of $(d-1)$ the variable Y^* satisfies arithmetic and harmonic mean bounds with the parameters ζ_1 and ζ_2 playing the role of volume fractions.

When nothing is known about the parameters ζ_1 and $\zeta_2 = 1 - \zeta_1$ we can still estimate Y^* using the inequality $0 \leq \zeta_1 \leq 1$. Assume $\sigma_1 \geq \sigma_2$ the bounds (2.21) imply

$$(d-1)\sigma_2 \leq Y^* \leq (d-1)\sigma_1, \tag{2.22}$$

which, when expressed in terms of σ^*, are the well-known Hashin-Shtrikman bounds on the effective conductivity of isotropic composites [5]. Consequently the Hashin-Shtrikman coated sphere geometries [5], (which attain these bounds) are examples of materials that correspond to the extreme values of ζ_1, namely $\zeta_1 = 0$ and $\zeta_1 = 1$.

The Beran bounds (2.21) become quite tight when the material is nearly homogeneous, i.e. when $(\sigma_1 - \sigma_2)$ is small. This is not surprising since the trial field more accurately represents the true field in the composite. To third-order in $(\sigma_1 - \sigma_2)$ the upper and lower bounds on σ^* coincide and imply

$$\begin{aligned}\sigma^* &= \langle\sigma\rangle - f_1 f_2(\sigma_1 - \sigma_2)^2/d\langle\sigma\rangle \\ &\quad + f_1 f_2(\sigma_1 - \sigma_2)^3(f_2 - f_1 + (d-1)(\zeta_1 - f_1))/d^2\langle\sigma\rangle + 0((\sigma_1 - \sigma_2)^3),\end{aligned} \tag{2.23}$$

where $\langle\sigma\rangle = f_1\sigma_1 + f_2\sigma_2$.

3. THE ANALYTIC METHOD

A different approach to bounding effective tensors was initiated by Bergman [6,7] and exploited further by Milton and McPhedran [8,9], Golden and Papanicolaou [10,11], Dell-Antonio, Orlandi and Figari [12], Dell-Antonio and Nesi [13] and others. One begins by considering the analytic properties of σ^* as a function of σ_1 and σ_2 keeping $\chi_1(\mathbf{x})$ fixed. Here we will suppose that $\sigma^*(\sigma_1,\sigma_2)$ is a scalar valued function that represents the effective conductivity of an isotropic two-phase composite. (It could also be chosen to represent a diagonal element of the effective conductivity tensor of an anisotropic composite).

The function $\sigma^*(\sigma_1,\sigma_2)$ is an analytic function of σ_1 and σ_2 except possibly at those points where the ratio σ_1/σ_2 is real and negative, zero or infinite. It satisfies the *Homogeneity Property*

$$\sigma^*(\lambda\sigma_1,\lambda\sigma_2)=\lambda\sigma^*(\sigma_1,\sigma_2) \ \ \forall\, \sigma, \tag{3.1}$$

and the *Energy-Dissipation Property*

$$\text{Re}\,\sigma^* > 0 \text{ whenever } \text{Re}\,\sigma_1 > 0 \text{ and } \text{Re}\,\sigma_2 > 0. \tag{3.2}$$

A mathematically rigorous proof of these properties has been given by Golden and Papanicolaou [10]. One immediate consequence of them is that

$$\text{Re}(\lambda\sigma^*) > 0 \tag{3.3}$$

for all complex λ such that $\text{Re}(\lambda\sigma_1) > 0$ and $\text{Re}(\lambda\sigma_2) > 0$. By letting λ range between these limits we deduce the *Wedge Property*

$$\sigma^*(\sigma_1,\sigma_2)\in W(\sigma_1,\sigma_2), \tag{3.4}$$

where $W(\sigma_1,\sigma_2)$, illustrated in figure 1, is the convex wedge in the complex plane bounded on one side by the ray from the origin passing through σ_1 and on the opposite side by the ray passing through σ_2. We will be studying bounds on the class of functions satisfying homogeneity and the wedge property. This class is clearly equivalent to the class of functions satisfying homogeneity and the energy dissipation property.

Figure 1: Example with complex component conductivities $\sigma_1 = -2+3i$ and $\sigma_2 = 1+i$ and volume fraction $f_1 = 0.6$ showing in the complex σ^* -plane the wedge W (dotted) and the region Ω' (shaded) that bound (i.e. contain) σ^*. The fractional linear transformation (3.17) maps Ω' onto the wedge W and thus preserves the basic analytic properties of the function.

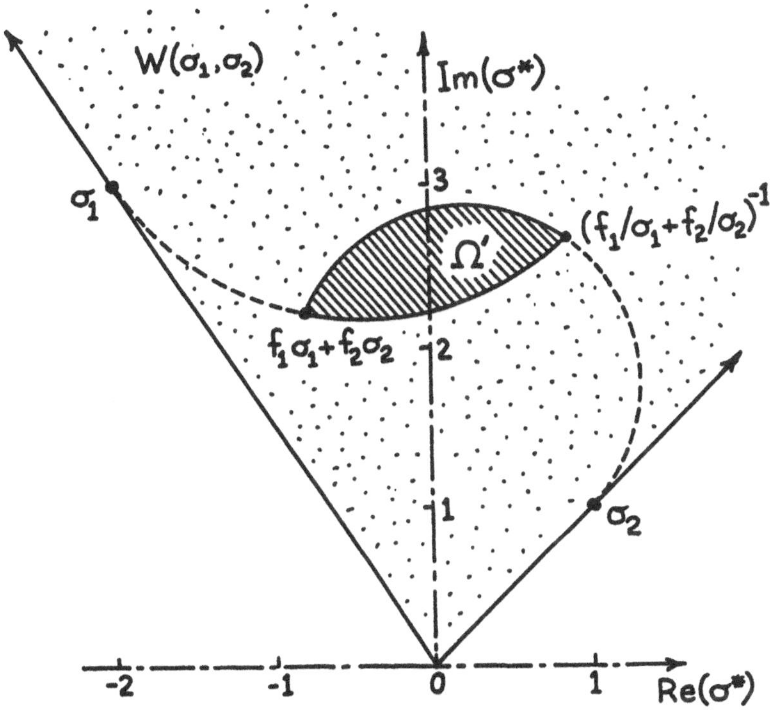

For simplicity (and without loss of generality: see [10]) let us only consider rational functions in this class. Any irrational function can be regarded as a limit of a sequence of rational functions. Since the functions are homogeneous we set $\sigma_2 = 1$. Next we ask the question: where are the poles and zeros of $\sigma^*(\sigma_1)$ located in the complex σ_1 -plane? The wedge property implies that $\mathrm{Im}\sigma^*$ and $\mathrm{Im}\sigma_1$ always take the same sign. Hence $\mathrm{Im}\sigma^*$ can only change sign across the real axis. Since $\mathrm{Im}\sigma^*$ changes sign 2n times as one moves once around an infinitesimal circle centered at a pole or zero order n it follows that all the poles and zeros are simple and located on the real axis. The residues associated with the poles must be real and negative to ensure that $\mathrm{Im}\sigma^*$ takes the right sign on each side of the real axis. This implies that the poles and zeros alternate along the real axis. Lastly they must be restricted to the negative real-axis (or zero or infinity) because the wedge property implies $\mathrm{Re}\sigma^* \geq 0$ along the positive real σ_1 axis.

In summary, the rational function $\sigma^*(\sigma_1) = \sigma^*(\sigma_1, 1)$ can be represented in the form

$$\sigma^*(\sigma_1) = k\frac{\prod_{i=1}^{m} (\sigma_1 - a_i)}{\prod_{i=1}^{m} (\sigma_1 - b_i)}, \tag{3.5}$$

where k is a positive constant and the zeros a_i and the poles b_i alternate along the negative real axis

$$0 \geq a_1 \geq b_1 \geq a_2 .... \geq a_m \geq b_m. \tag{3.6}$$

Equivalently, the function $\sigma^*(\sigma_1, \sigma_2) = \sigma_2 \sigma^*(\sigma_1/\sigma_2)$ can be represented in the more symmetric form

$$\sigma^*(\sigma_1, \sigma_2) = n\frac{\prod_{i=1}^{m+1} (p_i\sigma_1 + (1-p_i)\sigma_2)}{\prod_{i=1}^{m} (q_i\sigma_1 + (1-q_i)\sigma_2)}, \tag{3.7}$$

where n is a positive constant and

$$0 \leq p_1 \leq q_1 \leq p_2 \leq q_2 \cdots \leq q_m \leq p_{m+1} \leq 1. \tag{3.8}$$

To obtain useful bounds on σ^* we need additional information about the function. One source is the expansion (2.23). It can of course be derived without reference the variational bounds (see Brown [14]) and implies

$$\sigma^*(1,1) = n = 1, \tag{3.9}$$

$$\frac{\partial \sigma^*}{\partial \sigma_1} \Big|_{\sigma_1 = \sigma_2 = 1} = \sum_{i=1}^{m+1} p_i - \sum_{i=1}^{m} q_i = f_1. \tag{3.10}$$

To obtain bounds we now fix σ_1 and σ_2 and find the range of values taken by σ^* as the parameters $p_i, i = 1,2,...m+1$, and $q_i, i = 1,2,..m$ are varied subject to the constraints (3.8) and (3.10). Varying these parameters roughly corresponds to varying the geometry, although not all configurations of zeros and poles will correspond to real geometries. Since (3.10) can be used to express q_m in terms of the other parameters. We will regard q_m as a dependent parameter. If the constants p_i and q_i are varied by infinitesimal amounts δp_i and δq_i then from (3.7) the corresponding change in σ^* is

$$\begin{aligned} \delta\sigma^* &= \sigma^* \left[\sum_{i=1}^{m+1} \frac{\delta p_i}{p_i\sigma_1 + (1-p_i)\sigma_2} - \sum_{i=1}^{m} \frac{\delta q_i}{q_i\sigma_1 + (1-q_i)\sigma_2} \right] \\ &= \frac{\sigma^*(\sigma_1 - \sigma_2)}{q_m\sigma_1 + (1-q_m)\sigma_2} \left[\sum_{i=1}^{m+1} \alpha_i \delta p_i + \sum_{i=1}^{m-1} \beta_i \delta q_i \right] \end{aligned} \tag{3.11}$$

where

$$\alpha_i = \frac{(q_m - p_i)}{p_i\sigma_1 + (1-p_i)\sigma_2}, \qquad \beta_i = \frac{-(q_m - q_i)}{q_i\sigma_1 + (1-q_i)\sigma_2} \tag{3.12}$$

and where the constraint (3.10) has been used to eliminate δq_m (assuming $m \geq 1$)

Let's first suppose σ_1 and σ_2 are real. The variations δp_i and δq_i can independently be chosen to take either sign except when the set of p_i and q_i are such that one of the inequalities (3.8) becomes an equality. The possibilities that $p_i = q_i$ or that $p_i = q_{i-1}$ for some i can be ignored since these are equivalent to reducing the degree m of the rational function. So only δp_1 and δp_{m+1} are restricted: $\delta p_1 \geq 0$ when $p_1 = 0$ and $\delta p_{m+1} \leq 0$ when $p_{m+1} = 1$. When σ^* is at its maximum or minimum the coefficients α_i and β_i of the remaining unrestricted δp_i and δq_i in (3.11) must vanish. However, a quick inspection of (3.12) shows that these coefficients are not zero. To avoid a contradiction there cannot be any unrestricted p's or q's which we are free vary in both directions: i.e. either $m = 1, p_1 = 0, p_2 = 1$ and q_1 is determined by (3.10) or $m = 0$ and p_1 is determined by (3.10). These two possibilities correspond to the harmonic and arithmetic mean bounds. Thus we deduce

$$(f_1/\sigma_1 + f_2/\sigma_2)^{-1} \leq \sigma^* \leq f_1\sigma_1 + f_2\sigma_2. \tag{3.13}$$

One of the advantages of the analytic method is that it is easy is extend the method to produce bounds on σ^* when σ_1 and σ_2 are complex: this was first done, in independent work, by Bergman [7] and Milton [8]. We expect σ^* to range over a region $\Omega'(\sigma_1,\sigma_2)$ of the complex plane as the p_i and q_i are varied subject to (3.8) and (3.10). When σ^* is on the boundary of Ω' the coefficients of the independent unrestricted variations δp_i and δq_i need not vanish: instead they must lie along a line in the complex plane through the origin and parallel to the boundary of Ω' at σ^*, i.e. they must share a common phase. Otherwise, by an appropriate choice of δp_i or δq_i we could push σ^* outside Ω' which would contradict the definition of Ω'. Again an inspection of (3.12) shows that none of the coefficients α_i and β_i share a common phase since for any real constants a and b,

$$\arg\,(a\sigma_1 + (1-a)\sigma_2) \neq \arg\,(b\sigma_1 + (1-b)\sigma_2), \tag{3.14}$$

unless $a = b$ or $\mathrm{Im}(\sigma_1/\sigma_2) = 0$. To resolve this contradiction there can only be one unrestricted p_i or q_i when σ^* is on the boundary of Ω': i.e. $m = 1$ and either $p_1 = 0$ or $p_2 = 1$. Thus Ω' is bounded on one side by the circular are inscribed by

$$\sigma_\alpha^* = \sigma_2 + \frac{f_1\sigma_2(\sigma_1 - \sigma_2)}{\sigma_2 + \alpha f_2(\sigma_1 - \sigma_2)}, \tag{3.15}$$

as α varies from 0 to 1, and on the other side by the circular are inscribed by

$$\sigma_\beta^* = \sigma_1 + \frac{f_2\sigma_1(\sigma_2 - \sigma_1)}{\sigma_1 + \beta f_1(\sigma_2 - \sigma_1)}, \tag{3.16}$$

as β varies from 0 to 1. These circular arcs intersect at the arithmetic and harmonic means and, when extended into circles, pass through σ_2 and σ_1 respectively: see figure 1.

Now that these bounds are established it is easy to generate a whole hierarchy of bounds which include progressively more of the series expansion coefficients. The central idea [15] is to use these bounds on $\sigma^*(\sigma_1,\sigma_2)$ to define a new function $Y^*(\sigma_1,\sigma_2)$ satisfying the homogeneity and the wedge property. Since σ^* takes values in $\Omega'(\sigma_1,\sigma_2)$ and we want Y^* to take values in the wedge $W(\sigma_1,\sigma_2)$ we look for a transformation which maps the lens shaped region Ω' into the wedge W. A suitable choice of transformation is a fractional linear transformation which maps the corners of the lens, namely $\sigma^* = f_1\sigma_1 + f_2\sigma_2$ and $\sigma^* = (f_1/\sigma_1 + f_2/\sigma_2)^{-1}$ to infinity and zero. Such a transformation will map the circular arcs bounding Ω' into straight lines. Following this reasoning it quickly becomes apparent that the transformation

$$Y^* = \sigma_1\sigma_2(\sigma^*(f_1/\sigma_1 + f_2/\sigma_2) - 1)/(f_1\sigma_1 + f_2\sigma_2 - \sigma^*) \tag{3.17}$$

does the job: the arithmetic and harmonic means get mapped to infinity and zero and the points $\sigma^* = \sigma_1$ and $\sigma^* = \sigma_2$ which lie on extensions of the the circular arcs that bound Ω' get mapped to $Y^* = -\sigma_1$ and $Y^* = -\sigma_2$ which lie on extensions of the lines bounding W.

Simple algebraic manipulation of (3.17) shows that this definition of Y^* is equivalent to the previous definition (2.20). From (2.20) it is clear that the transformation to the variable Y^* has another important property. Any geometric parameter which first enters the series expansion of σ^* in powers of $(\sigma_1 - \sigma_2)$ at the kth order will enter the series expansion of Y^* at the (k-2)th order, i.e. the transformation shifts the coefficients in the series expansion to a lower order. From (2.23) the expansion of Y^* to first order in $\sigma_1 - \sigma_2$ is

$$Y^* = (d-1)(\zeta_1\sigma_1 + \zeta_2\sigma_2) + 0(\sigma_1 - \sigma_2)^2. \tag{3.18}$$

Ideally we want a function with properties as close to those of a conductivity function as possible. Therefore we normalize $Y^*(\sigma_1,\sigma_2)$ to obtain the function

$$\sigma'^*(\sigma_1,\sigma_2) = Y^*(\sigma_1,\sigma_2)/(d-1), \tag{3.19}$$

which satisfies homogeneity, the wedge property, and

$$\sigma'^*(1,1) = 1, \tag{3.20}$$

$$\frac{\partial\sigma'^*}{\partial\sigma_1}\Big|_{\sigma_1=\sigma_2=1} = \zeta_1. \tag{3.21}$$

Using these properties we easily deduce the Beran bounds (3.21) using the same line of reasoning as led to the arithmetic and harmonic mean bounds (3.13).

Next we can introduce the function

$$Y'^* = \sigma_1\sigma_2(\sigma'^*(\zeta_1/\sigma_1 + \zeta_2/\sigma_2) - 1)/(\zeta_1\sigma_1 + \zeta_2\sigma_2 - \sigma'^*), \tag{3.22}$$

which satisfies the homogeneity and the wedge property. Continuing in this way we generate a whole hierarchy of nested bounds incorporating progressively more coefficients of the series expansion of σ^* in powers of $\sigma_1 - \sigma_2$. When σ_1 and σ_2 complex these bounds form nested sequence of lens-shaped regions in the complex plane [8,9]. For isotropic composites in two dimensions Keller [16] and Dykhne [17] proved the reciprocal relation

$$\sigma^*(1/\sigma_1,1/\sigma_2)=1/\sigma^*(\sigma_1,\sigma_2). \tag{3.23}$$

One interesting feature of the transformation to the function $\sigma'^*(\sigma_1,\sigma_2)$ is that the reciprocal relation is preserved. Indeed it follows from (3.17) that when $d=2$

$$\sigma'^*(1/\sigma_1,1/\sigma_2)=1/\sigma'^*(\sigma_1,\sigma_2). \tag{3.24}$$

4. THE TRANSLATION METHOD

The simplest bounds are of course the arithmetic and harmonic mean bounds. The translation method, developed by Tartar and Murat [18,19] and Lurie and Cherkaev [20,21], applies these simple bounds to the effective tensor of a comparison medium, obtained by translating the tensor field of the original medium. To illustrate the method (see also [22], [23]) we will derive the Hashin-Shtrikman lower bound on the effective bulk modulus κ^*: this was first accomplished using the translation method by Francfort and Murat [24].

Consider a two phase isotropic medium with elasticity tensor

$$\mathbf{C}(\mathbf{x})=\mathbf{C}_1\chi_1(\mathbf{x})+\mathbf{C}_2\chi_2(\mathbf{x}), \tag{4.1}$$

where for $\alpha=1,2$

$$(C_\alpha)_{ijkl}=\kappa_\alpha\delta_{ij}\delta_{kl}+\mu_\alpha(\delta_{ik}\delta_{jl}+\delta_{il}\delta_{jk}-(2/d)\delta_{ij}\delta_{kl}) \tag{4.2}$$

and κ_1,κ_2 and μ_1,μ_2 are the bulk and shear moduli of the two phases. The comparison medium has elasticity tensor

$$\mathbf{C}'(\mathbf{x})=\mathbf{C}(\mathbf{x})-\boldsymbol{\tau}, \tag{4.3}$$

where the constant translation tensor $\boldsymbol{\tau}$ is chosen so that

(i) the comparison medium is physically reasonable, i.e. $\mathbf{C}'(\mathbf{x})$ is positive semidefinite

(ii) the quadratic form associated with $\boldsymbol{\tau}$ is quasiconvex, i.e. convex on the subspace of strain fields. The requirement of quasiconvexity is satisfied if and only if the inequality

$$<\varepsilon^T:\tau\varepsilon> \geq <\varepsilon^T>:\tau<\varepsilon> \tag{4.4}$$

(where : denotes a double contraction of indices) holds for all periodic strain fields

$$\varepsilon(\mathbf{x})=\frac{1}{2}(\nabla\mathbf{u}(\mathbf{x})+(\nabla\mathbf{u}(\mathbf{x}))^T)\equiv\overline{\nabla\mathbf{u}(\mathbf{x})}. \tag{4.5}$$

It ensures that the effective tensors of the two media satisfy the simple inequality

$$\mathbf{C}'^{*} \leq \mathbf{C}^{*} - \boldsymbol{\tau}. \tag{4.6}$$

This inequality is proved by substituting (4.4) in the variational principle

$$\boldsymbol{\varepsilon}^{*T}:\mathbf{C}'^{*}\boldsymbol{\varepsilon}^{*} = \min_{\substack{\boldsymbol{\varepsilon}(\mathbf{x}) = \nabla \mathbf{u}(\mathbf{x}) \\ <\boldsymbol{\varepsilon}(\mathbf{x})> = \boldsymbol{\varepsilon}^{*}}} < \boldsymbol{\varepsilon}(\mathbf{x}):\mathbf{C}(\mathbf{x})\boldsymbol{\varepsilon}(\mathbf{x}) > - < \boldsymbol{\varepsilon}(\mathbf{x}):\boldsymbol{\tau}\boldsymbol{\varepsilon}(\mathbf{x})>. \tag{4.7}$$

From the harmonic mean bounds

$$(\mathbf{C}'^{*})^{-1} \leq f_1(\mathbf{C}_1 - \boldsymbol{\tau}) + f_2(\mathbf{C}_2 - \boldsymbol{\tau}), \tag{4.8}$$

and (4.6) we deduce the basic inequality

$$(\mathbf{C}^{*} - \boldsymbol{\tau})^{-1} \leq f_1(\mathbf{C}_1 - \boldsymbol{\tau})^{-1} + f_2(\mathbf{C}_2 - \boldsymbol{\tau})^{-1} \tag{4.9}$$

of the translation method.

Let us select an isotropic translation operator

$$\tau_{ijkl} = \kappa_0 \delta_{ij}\delta_{kl} + \mu_0(\delta_{ik}\delta_{jl} + \delta_{il}\delta_{jl} - (2/d)\delta_{ij}\delta_{kl}). \tag{4.10}$$

It follows from (4.4) that the quadratic form associated with $\boldsymbol{\tau}$ will be quasiconvex if and only if for all periodic strains $\boldsymbol{\varepsilon}(\mathbf{x})$,

$$\sum_{\mathbf{k}\neq 0} \hat{\boldsymbol{\varepsilon}}(\mathbf{k}):\boldsymbol{\tau}\hat{\boldsymbol{\varepsilon}}(-\mathbf{k}) \geq 0, \tag{4.11}$$

where

$$\hat{\boldsymbol{\varepsilon}}(\mathbf{k})_{ij} = \frac{1}{2}(k_i\hat{u}(\mathbf{k})_j + k_j\hat{u}(\mathbf{k})_i) \tag{4.12}$$

is the Fourier transform of $\boldsymbol{\varepsilon}(\mathbf{x})$ and $\hat{\mathbf{u}}(\mathbf{k})$ is the Fourier transform of the associated displacement field. Since the Fourier components $\hat{\mathbf{u}}(\mathbf{k})$ can be chosen independently $\boldsymbol{\tau}$ is quasiconvex if and only if each term in the sum (4.11) is positive. By substituting (4.12) in (4.11) and using Schwartz's inequality we deduce that τ is quasiconvex if and only if

$$\mu_0 \geq 0, \quad \kappa_0 \geq -2(d-1)\mu_0/d. \tag{4.13}$$

Setting

$$\kappa_0 = -2(d-1)\mu_0/d, \qquad \mu_0 = \min\{\mu_1, \mu_2\}. \tag{4.14}$$

ensures both quasiconvexity and that $\mathbf{C}'(\mathbf{x})$ is positive semidefinite. The resulting bound

$$(\kappa^* - \kappa_0)^{-1} \le f_1(\kappa_1 - \kappa_0)^{-1} + f_2(\kappa_2 - \kappa_0)^{-1} \tag{4.15}$$

implied by (4.9) is in fact the well-known Hashin-Shtrikman lower bound on the effective bulk modulus [25]. The corresponding upper bound can be derived by translating the compliance tensor $(\mathbf{C}(\mathbf{x}))^{-1}$ and applying the arithmetic mean bound to the effective elasticity tensor of the comparison problem. The derivation of the Hashin-Shtrikman shear modulus bounds using the translation bounds is not as simple but can be achieved by imbedding the fourth-order elasticity problem in a problem involving eighth-order tensors: see [23,26] for details.

The basic tensor inequality (4.9) of the translation method can be rewritten as

$$\mathbf{C}^* \ge \boldsymbol{\tau} + (f_1(\mathbf{C}_1 - \boldsymbol{\tau})^{-1} + f_2(\mathbf{C}_2 - \boldsymbol{\tau})^{-1})^{-1}. \tag{4.16}$$

To reduce this to a form where $\boldsymbol{\tau}$ appears only once we note that for any matrix $\mathbf{X}$ and constants α and β satisfying $\alpha + \beta = 1$ we have

$$\begin{aligned}
&\alpha\mathbf{X} + \beta\mathbf{I} - \alpha\beta(\mathbf{X} - \mathbf{I})(\alpha\mathbf{I} + \beta\mathbf{X})^{-1}(\mathbf{X} - \mathbf{I}) \\
&\quad = \left[(\alpha\mathbf{X} + \beta\mathbf{I})(\alpha\mathbf{I} + \beta\mathbf{X}) - \alpha\beta(\mathbf{X} - \mathbf{I})^2\right](\alpha\mathbf{I} + \beta\mathbf{X})^{-1} \\
&\quad = (\alpha + \beta)^2\mathbf{X}(\alpha\mathbf{I} + \beta\mathbf{X})^{-1} \\
&\quad = (\alpha\mathbf{X}^{-1} + \beta\mathbf{I})^{-1}.
\end{aligned} \tag{4.17}$$

Setting $\alpha = f_1, \beta = f_2$ and $\mathbf{X} = (\mathbf{C}_1 - \boldsymbol{\tau})(\mathbf{C}_2 - \boldsymbol{\tau})^{-1}$ and multiplying on the right by $\mathbf{C}_2 - \boldsymbol{\tau}$ produces

$$\begin{aligned}
&(f_1(\mathbf{C}_1 - \boldsymbol{\tau})^{-1} + f_2(\mathbf{C}_2 - \boldsymbol{\tau})^{-1})^{-1} \\
&\quad = f_1\mathbf{C}_1 + f_2\mathbf{C}_2 - \boldsymbol{\tau} - f_1 f_2(\mathbf{C}_1 - \mathbf{C}_2)(f_2\mathbf{C}_1 + f_1\mathbf{C}_2 - \boldsymbol{\tau})^{-1}(\mathbf{C}_1 - \mathbf{C}_2).
\end{aligned} \tag{4.18}$$

Substituting this in (4.16) and assuming $\mathbf{C}_1 - \mathbf{C}_2$ is non-singular gives the bound

$$\mathbf{Y}^* + \boldsymbol{\tau} \ge 0, \tag{4.19}$$

where $\mathbf{Y}^*$ is defined, by analogy with (2.20), via

$$\mathbf{C}^* = f_1\mathbf{C}_1 + f_2\mathbf{C}_2 - f_1 f_2(\mathbf{C}_1 - \mathbf{C}_2)(f_1\mathbf{C}_2 + f_2\mathbf{C}_1 + \mathbf{Y}^*)^{-1}(\mathbf{C}_1 - \mathbf{C}_2). \tag{4.20}$$

This surprisingly simple form (4.19) of the translation bounds in terms of the tensor $\mathbf{Y}^*$ was first noted by Cherkaev and Gibiansky [1]. To obtain their result they assumed that $\mathbf{C}_1, \mathbf{C}_2$

and $\mathbf{C}^*$ commute. We have shown this assumption is unnecessary when $\mathbf{Y}^*$ is defined according to (4.20).

5. THE EFFECTIVE MEDIUM APPROXIMATION

One of the best known approximations for estimating the effective conductivity of an isotropic two-component granular aggregate is Bruggeman's effective medium approximation

$$\frac{f_1(\sigma_1-\sigma^*)}{\sigma_1+(d-1)\sigma^*}+\frac{f_2(\sigma_2-\sigma^*)}{\sigma_2+(d-1)\sigma^*}=0, \tag{5.1}$$

that applies to a space-filling aggregate of grains, treated as spherical in the simplest approximation. The approximation is realizable [27,28] and corresponds exactly to a hierarchical model with a wide distribution of sphere sizes. The distribution is sufficiently wide that any pair of spheres of comparable size are well separated from each other: the intervening space is filled with spheres of much smaller or much larger size.

Berryman [29] noted that the effective medium approximation takes a simple form when the variable Y^* defined by (2.20) is introduced. From (5.1) we have

$$f_1-\frac{df_1\sigma^*}{\sigma_1+(d-1)\sigma^*}+f_2-\frac{df_2\sigma^*}{\sigma_2+(d-1)\sigma^*}=0, \tag{5.2}$$

which can be rewritten as

$$\sigma^*=-(d-1)\sigma^*+(f_1/(\sigma_1+(d-1)\sigma^*)+f_2/(\sigma_2+(d-1)\sigma^*))^{-1}, \tag{5.3}$$

Comparing this with (4.16) and recalling its equivalent form (4.19) we see that the effective medium approximation is equivalent to the equation

$$Y^*=(d-1)\sigma^*. \tag{5.4}$$

6. THE FIELD EQUATION RECURSION METHOD

The importance of the tensor $\mathbf{Y}^*$ is now evident but we are clearly lacking a physical interpretation of this tensor. The field equation recursion method, developed in [30] provides such an interpretation. The basic idea of the method is to define a hierarchy of "effective tensors" associated with a nested sequence of Hilbert spaces each containing progressively fewer fields. The nesting of the Hilbert spaces links their "effective tensors" and the basic premise of the method is that crude bounds on any "effective tensor" in this hierarchy will provide narrow bounds on the original effective tensor, σ^* or $\mathbf{C}^*$, of interest. The tensor $\mathbf{Y}^*$ is the first such "effective tensor" that appears in this hierarchy after σ^*, and its associated Hilbert space is obtained by stripping uniform fields from the original Hilbert space of square integrable periodic fields.

The driving fields are component-wise constant, average zero, fields of the form

$$\mathbf{p}^*(\mathbf{x}) = f_2\chi_1(\mathbf{x})\mathbf{u} - f_1\chi_2(\mathbf{x})\mathbf{u}, \tag{6.1}$$

where $\mathbf{u}$ is a vector for the conductivity problem or a symmetric tensor for the elasticity problem. The tensor $\mathbf{Y}^*$ acts locally and is independent of $\mathbf{x}$ and is defined via

$$<\mathbf{p}^*(\mathbf{x})\cdot\mathbf{Y}^*\mathbf{p}^*(\mathbf{x})> = \min_{\substack{\mathbf{P} \\ <\chi_i\mathbf{P}>=0 \\ \nabla\times(\mathbf{p}^*+\mathbf{P})=0}} <\mathbf{P}(\mathbf{x})\cdot\sigma(\mathbf{x})\,\mathbf{P}(\mathbf{x})>, \tag{6.2}$$

for the conductivity problem and via

$$<\mathbf{p}^*(\mathbf{x})^T:\mathbf{Y}^*\mathbf{p}^*(\mathbf{x})> = \min_{\substack{\mathbf{P} \\ <\chi_i\mathbf{P}>=0 \\ \mathbf{p}^*+\mathbf{P}=\overline{\nabla\mathbf{u}}}} <\mathbf{P}(\mathbf{x})^T:\mathbf{C}(\mathbf{x})\mathbf{P}(\mathbf{x})>, \tag{6.3}$$

for the elasticity problem. It is easy to link this tensor with the effective conductivity tensor (or effective elasticity tensor). From the variational definition (2.1) of σ^* we have

$$\begin{aligned}
\mathbf{e}^*\cdot\sigma^*\mathbf{e}^* &= \min_{\substack{\mathbf{p}^*,\mathbf{P} \\ <\chi_i\mathbf{P}>=0 \\ \nabla\times(\mathbf{p}^*+\mathbf{P})=0}} <(\mathbf{e}^*+\mathbf{p}^*+\mathbf{P})\cdot\sigma(\mathbf{x})\,(\mathbf{e}^*+\mathbf{p}+\mathbf{P})> \\
&= \min_{\mathbf{p}^*} <(\mathbf{e}^*+\mathbf{p}^*)\cdot\sigma(\mathbf{x})(\mathbf{e}^*+\mathbf{p}^*)> + <\mathbf{p}^*\cdot\mathbf{Y}^*\mathbf{p}^*> \\
&= \mathbf{e}^*\cdot<\sigma>\mathbf{e}^* + f_1f_2\min_{\mathbf{u}}\,(\mathbf{u}\cdot(\mathbf{Y}^*+f_2\sigma_1+f_1\sigma_2)\mathbf{u} + 2\mathbf{u}\cdot(\sigma_1-\sigma_2)\mathbf{e}^*) \\
&= \mathbf{e}^*\cdot(f_1\sigma_1+f_2\sigma_2-f_1f_2(\sigma_1-\sigma_2)(f_1\sigma_2+f_2\sigma_1+\mathbf{Y}^*)^{-1}(\sigma_1-\sigma_2))\mathbf{e}^*,
\end{aligned} \tag{6.4}$$

which is consistent with (2.20) and (4.20).

The tensor $\mathbf{Y}^*$ can also be defined in terms of a dual variational principle. The driving fields are again piecewise constant, average zero, fields of the form

$$\mathbf{q}^*(\mathbf{x}) = f_2\chi_1(\mathbf{x})\mathbf{v} - f_1\chi_2(\mathbf{x})\mathbf{v} \tag{6.5}$$

where $\mathbf{v}$ is a constant field, and the tensor $\mathbf{Y}^*$ is defined for the conductivity problem via the dual principle

$$<\mathbf{q}^*(\mathbf{x})\cdot(\mathbf{Y}^*)^{-1}\mathbf{q}^*(\mathbf{x})> \; = \min_{\substack{\mathbf{Q} \\ <\chi_i\mathbf{Q}> = 0 \\ \nabla\cdot(\mathbf{q}^* + \mathbf{Q}) = 0}} \; <\mathbf{Q}(\mathbf{x})\cdot\sigma(\mathbf{x})^{-1}\mathbf{Q}(\mathbf{x})> \tag{6.6}$$

The equivalence of the two definitions (6.2) and (6.6) is easily proved using standard variational techniques [22]. If $\mathbf{P}(\mathbf{x})$ is the minimizer of (6.2) associated with the driving field $\mathbf{p}^*(\mathbf{x})$ then

$$\mathbf{Q}(\mathbf{x}) = \sigma(\mathbf{x})\mathbf{P}(\mathbf{x}) \tag{6.7}$$

is the minimizer of the dual principle (6.6) associated with the driving field

$$\mathbf{q}^*(\mathbf{x}) = -\mathbf{Y}^*\mathbf{p}^*(\mathbf{x}) \tag{6.8}$$

The minus sign in (6.8) is needed to ensure that

$$\begin{aligned} <\mathbf{p}^*(\mathbf{x})\cdot\mathbf{Y}^*\mathbf{p}^*(\mathbf{x})> &= -<\mathbf{p}^*(\mathbf{x})\cdot\mathbf{q}^*(\mathbf{x})> \\ &= <\mathbf{P}(\mathbf{x})\cdot\mathbf{Q}(\mathbf{x})> - <(\mathbf{p}^* + \mathbf{P})\cdot(\mathbf{q}^* + \mathbf{Q})> \\ &= <\mathbf{P}(\mathbf{x})\cdot\mathbf{Q}(\mathbf{x})> = <\mathbf{P}(\mathbf{x})\cdot\sigma(\mathbf{x})\mathbf{P}(\mathbf{x})> \end{aligned} \tag{6.9}$$

in accordance with (6.2). (It follows from $\nabla\cdot(\mathbf{q}^* + \mathbf{Q}) = 0$ and $\nabla\times(\mathbf{p}^* + \mathbf{P}) = 0$ that $<(\mathbf{p}^* + \mathbf{P})\cdot(\mathbf{q}^* + \mathbf{Q})> \; = 0$).

The translation bounds (4.19) also follow directly from the definition (6.2) of $\mathbf{Y}^*$. Quasiconvexity implies

$$<(\mathbf{p}^* + \mathbf{P})\cdot\tau(\mathbf{p}^* + \mathbf{P})> = <\mathbf{p}^*\cdot\tau\mathbf{p}^*> + <\mathbf{P}\cdot\tau\mathbf{P}> \geq 0 \tag{6.10}$$

and because $\mathbf{C}(\mathbf{x}) \geq \tau$ we have from (6.3)

$$<\mathbf{p}^*\cdot\mathbf{Y}^*\mathbf{p}^*> \; \geq \min_{\substack{\mathbf{P} \\ <\chi_i\mathbf{P}> = 0 \\ \mathbf{p}^* + \mathbf{P} = \nabla u}} \; <\mathbf{P}\cdot\tau\mathbf{P}> \; \geq -<\mathbf{p}^*\cdot\tau\mathbf{p}^*> \tag{6.11}$$

which implies $\mathbf{Y}^* + \boldsymbol{\tau} \geq 0$.

7. THE EXTENSION TO MULTICOMPONENT COMPOSITES

The definition (6.2) of $\mathbf{Y}^*$ also applies to n-component composites comprised of possibly anisotropic phases. However $\mathbf{Y}^*$ changes its character from a local constant operator to a non-local operator which maps component-wise constant mean zero fields to component-wise constant mean zero fields. To represent its action we select one of the components, say component q, as a reference component and introduce an orthonormal basis $\mathbf{e}_1, \mathbf{e}_2, \ldots, \mathbf{e}_d$ of uniform fields (for elasticity there would be $d(d+1)/2$ such fields) and say the basis

$$\mathbf{v}_{bl} = f_q \chi_b \mathbf{e}_l - f_b \chi_q \mathbf{e}_l \tag{7.1}$$

of component-wise constant mean zero fields: here the component index b ranges from 1 to n, excluding $b = q$, and the direction index l ranges from 1 to d (or $d(d+1)/2$). Other choices of basis fields are equally or perhaps more appropriate: for example in [30] the basis fields are chosen to be orthonormal. The operators $\boldsymbol{\sigma}^*$ and $\mathbf{Y}^*$ are self-adjoint and represented by matrices $\underline{\underline{\sigma}}^*$ and $\underline{\underline{Y}}^*$ with components determined by the action of the operators on the basis fields:

$$\boldsymbol{\sigma}^* \mathbf{e}_l = \sum_k \mathbf{e}_k \sigma^*_{kl} \;, \quad \mathbf{Y}^* \mathbf{v}_{bl} = \sum_{\substack{a,k \\ a \neq q}} \mathbf{v}_{ak} \, Y^*_{ak,bl} \tag{7.2}$$

Thus $\mathbf{Y}^*$ is represented by a $d(n-1) \times d(n-1)$ matrix (or by a $d(d+1)(n-1)/2 \times d(d+1)\,(n-1)/2$ matrix for elasticity). The driving fields $\mathbf{e}^*$ and $\mathbf{p}^*$ are each expanded in the basis fields,

$$\mathbf{e}^* = \sum_l \beta_l \mathbf{e}_l \;, \quad \mathbf{p}^* = \sum_{\substack{b,l \\ b \neq q}} \alpha_{bl} \mathbf{v}_{bl} \tag{7.3}$$

and we have

$$\langle \mathbf{e}^* \cdot \boldsymbol{\sigma} \mathbf{e}^* \rangle = \sum_{\substack{a,l,k \\ a \neq q}} \beta_k \beta_l f_a \sigma^a_{kl},$$

$$\langle \mathbf{e}^* \cdot \boldsymbol{\sigma} \mathbf{p}^* \rangle = \sum_{\substack{a,l,k \\ a \neq q}} \alpha_{ak} \, \beta_l \, f_a f_q (\sigma^a_{kl} - \sigma^q_{kl}),$$

$$<\mathbf{p}^*\cdot\sigma\mathbf{p}^*> = \sum_{\substack{a,l,b,k\\ a,b\neq q}} \alpha_{ak}\alpha_{bl}\,(\delta_{ab}\, f_a(f_q)^2\sigma^a_{kl} + f_a f_b f_q \sigma^q_{kl}),$$

$$<\mathbf{p}^*\cdot\mathbf{Y}^*\mathbf{p}^*> \;=\; \sum_{\substack{a,l,b,k,c\\ a,b,c\neq q}} \alpha_{ak}\alpha_{bl}\, Y^*_{ck,bl}\,(\delta_{ac} f_a(f_q)^2 + f_a f_c f_q), \tag{7.4}$$

where σ^a_{lk} represents the matrix associated with the conductivity tensor in component a. These results imply

$$\mathbf{e}^*\cdot\sigma^*\mathbf{e}^* = \sum_{k,l} \beta_k\sigma^*_{kl}\beta_l = \underline{\beta}\cdot\underline{\underline{\sigma}}^*\underline{\beta}$$

$$= \min_{\substack{\mathbf{p}^*,\mathbf{P}\\ <\chi_i\mathbf{P}>=0\\ \nabla\times(\mathbf{p}^*+\mathbf{P})=0}} \;<(\mathbf{e}^* + \mathbf{p}^* + \mathbf{P})\cdot\sigma(\mathbf{e}^* + \mathbf{p}^* + \mathbf{P})>$$

$$= \min_{\mathbf{p}^*} \;<(\mathbf{e}^* + \mathbf{p}^*)\cdot\sigma(\mathbf{e}^* + \mathbf{p}^*)> + <\mathbf{p}^*\cdot\mathbf{Y}^*\mathbf{p}^*>$$

$$= \min_{\underline{\alpha}} \,(\underline{\beta}\cdot\sum_a f_a\underline{\underline{\sigma}}^a\,\underline{\beta} + 2\,\underline{\alpha}\cdot\underline{\underline{L}}\,\underline{\beta} + \underline{\alpha}\cdot\underline{\underline{M}}\,\underline{\alpha}) \tag{7.5}$$

where $\underline{\underline{L}}$ is the $d(n-1)\times d$ rectangular matrix with components

$$L_{bk,l} = f_b f_q(\sigma^b_{kl} - \sigma^q_{kl}) \tag{7.6}$$

and $\underline{\underline{M}}$ is the $d(n-1)\times d(n-1)$ square matrix with components

$$M_{ak,bl} = \delta_{ab} f_a(f_q)^2\sigma^b_{kl} + f_a f_b f_q \sigma^q_{kl}$$
$$+ \sum_{c\neq q} Y^*_{ck,bl}(\delta_{ac} f_a(f_q)^2 + f_a f_c f_q) \tag{7.7}$$

The self-adjointness of $\mathbf{Y}^*$ implies $\underline{\underline{M}}$ is a symmetric matrix. (However $\underline{\underline{Y}}$ itself is not symmetric because the basis fields (7.1) are not mutually orthogonal). Taking the minimum over $\underline{\alpha}$ in (7.5) gives the desired equation

$$\underline{\underline{\sigma}}^* = \sum_a f_a\underline{\underline{\sigma}}^a - \underline{\underline{L}}^T\underline{\underline{M}}^{-1}\underline{\underline{L}} \tag{7.8}$$

linking the matrix representing σ^* with the matrix representing $\mathbf{Y}^*$, where $\underline{\underline{L}}$ and $\underline{\underline{M}}$ are given by (7.6) and (7.7).

When the quadratic form associated with an operator $\boldsymbol{\tau}$ is quasiconvex and $\boldsymbol{\sigma}(\mathbf{x}) \geq \boldsymbol{\tau}$ for all $\mathbf{x}$ (where $\boldsymbol{\sigma}(\mathbf{x})$ may represent the elasticity tensor $\mathbf{C}(\mathbf{x})$ or some other tensor field) then the derivation (6.6) of the inequality

$$<\mathbf{p}^*\cdot\mathbf{Y}^*\mathbf{p}^* > \geq - <\mathbf{p}^*\cdot\boldsymbol{\tau}\mathbf{p}^* > \tag{7.9}$$

extends to multicomponent composites. We now establish that (7.9) implies the translation bounds for multicomponent media. The operator $\boldsymbol{\tau}$ acting on component-wise mean zero fields is represented by a matrix $\tau_{ak,bl} = \delta_{ab}\tau_{kl}$. Hence (7.9) produces

$$\underline{\underline{M}} \geq \underline{\underline{M}}'$$

where the matrix $\underline{\underline{M}}'$ has components

$$M'_{ak,bl} = \delta_{ab} f_a (f_q)^2 (\sigma^a_{kl} - \tau_{kl}) + f_a f_b f_q (\sigma^q_{kl} - \tau_{kl}) \tag{7.10}$$

and is positive semidefinite. Notice that $\underline{\underline{M}}'$ only incorporates $\underline{\underline{\sigma}}^a, \underline{\underline{\sigma}}^q$ and $\underline{\underline{\tau}}$ in the combinations $\underline{\underline{\sigma}}^a - \underline{\underline{\tau}}$ and $\underline{\underline{\sigma}}^q - \underline{\underline{\tau}}$. Moreover we have

$$<\mathbf{e}^*\cdot<(\boldsymbol{\sigma}(\mathbf{x}) - \boldsymbol{\tau})^{-1} >^{-1}\mathbf{e}^* > = \min_{\mathbf{p}^*} < (\mathbf{e}^* + \mathbf{p}^*)\cdot(\boldsymbol{\sigma}(\mathbf{x}) - \boldsymbol{\tau})(\mathbf{e}^* + \mathbf{p}^*) >$$

$$= \sum_a f_a \underline{\beta}^T (\underline{\underline{\sigma}}^a - \underline{\underline{\tau}})\underline{\beta} - \underline{\beta}^T \underline{\underline{L}}^T (\underline{\underline{M}}')^{-1} \underline{\underline{L}}\underline{\beta} \tag{7.11}$$

From a comparison of (7.8) and (7.11) it follows that the operator inequality $\mathbf{Y}^* + \boldsymbol{\tau} \geq 0$ (or equivalently the matrix inequality $\underline{\underline{M}} \geq \underline{\underline{M}}' \geq 0$) implies the bound

$$(\boldsymbol{\sigma}^* - \boldsymbol{\tau})^{-1} \leq < (\boldsymbol{\sigma}(\mathbf{x}) - \boldsymbol{\tau})^{-1} > \tag{7.12}$$

of the translation method.

One important conclusion follows from this analysis. From the translation bounds (7.12) it is not clear how far we should translate $\boldsymbol{\sigma}$ to produce the best bound, i.e. if we replace $\boldsymbol{\tau}$ by $c\boldsymbol{\tau}$ then it is not clear what is the best choice for c. However from the bounds (7.9) which are linear in $\boldsymbol{\tau}$ it is clear that we should translate as far as possible, i.e. until $\boldsymbol{\sigma}(\mathbf{x}) - \boldsymbol{\tau}$ becomes singular, or not at all. A similar conclusion is stated in Section 8 of [26]. However the reasoning given there is incorrect as noted by Allaire and Kohn [31].

Acknowledgements

I am grateful to G. Allaire, J. Berryman, A.V. Cherkaev, G. Francfort, L. Gibiansky, K. Golden, R.V. Kohn and G. Papanicolaou for discussions which motivated this article. D. Rosenthal is thanked for her care in preparing the manuscript. Support from a Packard fellowship, a Sloan fellowship, the Air Force Office of Scientific Research and the Army Research Office is gratefully acknowledged.

REFERENCES

[1] L.V. Gibiansky and A.V. Cherkaev, private communication, Dijon, 1989.

[2] M.J. Beran, "Use of the variational approach to determine bounds for the effective permittivity in random media," Nuovo Cimento 38 (1965), 771-782.

[3] G.W. Milton, "Bounds on the electromagnetic, elastic, and other properties of two-component composites," Phys. Rev. Lett. 46 (1981) 542-545.

[4] S. Torquato and G. Stell, "Bounds on the effective thermal conductivity of a dispersion of fully penetrable spheres," Lett. Appl. Eng. Sci., 23 (1985), 375-384.

[5] Z. Hashin and S. Shtrikman, "A variational approach to the theory of the effective magnetic permeability of multiphase materials," J. Appl. Phys. 33 (1962) 1514-1517.

[6] D.J. Bergman, "The dielectric constant of a composite material-a problem in classical physics," Phys. Rep. C43 (1978), 377-407.

[7] D.J. Bergman, "Rigorous bounds for the complex dielectric constant of a two-component composite," Ann. Phys. 138 (1982), 78-114.

[8] G.W. Milton, "Bounds on the complex permittivity of a two-component composite material," J. Appl. Phys. 52 (1981), 5286-5293; "Bounds on the transport and optical properties of a two-component composite material," J. Appl. Phys. 52 (1981), 5294-5304.

[9] R.C. McPhedran and G.W. Milton, "Bounds and exact theories for the transport properties of inhomogeneous media," Appl. Phys. A26 (1981), 207-220.

[10] K. Golden and G. Papanicolaou, "Bounds for effective parameters of heterogeneous media by analytic continuation," Commun. Math. Phys. 90 (1983) 473.

[11] K. Golden and G. Papanicolaou, "Bounds for effective parameters of multicomponent media by analytic continuation," J. Stat. Phys. 40 (1985) 655.

[12] G.F. Dell' Antonio, R. Figari, and E. Orlandi, "An approach through orthogonal projections to the study of inhomogeneous or random media with linear response," Ann. Inst. Henri Poincare 44 (1986) 1-28.

[13] G.F. Dell' Antonio and V. Nesi, "A general representation for the effective dielectric constant of a composite," J. Math. Phys. 29 (1988) 2688-2694.

[14] W.F. Brown, "Solid mixture permittivities," J. Chem. Phys. 23 (1955), 1514-1517.

[15] G.W. Milton and K. Golden, "Thermal conduction in composites," in *Thermal Conductivity* 18, pp. 571-582, ed. by T. Ashworth and D.R. Smith, Plenum, 1985.

[16] J.B. Keller, "A theorem on the conductivity of a composite medium," J. Math. Phys. 5 (1964) 548-549.

[17] A.M. Dykhne, "Conductivity of a two-dimensional, two-phase system," Zh. Eksp. Teor. Fiz 59 (1970) 110-115 [Soviet Phys JETP 32 (1971) 63-65].

[18] L. Tartar, "Estimations fines des coefficients homogeneises," in *Ennio De Giorgi's Colloquium,* ed. P. Kree, Research notes in mathematics 125, pp. 168-187, Pitman Press, London, 1985.

[19] F. Murat and L. Tartar, "Calcul des variations et homogeneisation," in *Les Methodes d'Homogeneisation: Theorie et Applications en Physique,* Coll. de la Dir. des Etudes et Recherches d'Electricite de France, pp. 319-369 Eyrolles, Paris, 1985.

[20] K.A. Lurie and A.V. Cherkaev, "Exact estimates of the conductivity of a binary mixture of isotropic components," Proc. Roy. Soc. Edinburgh 104A (1986) 21-38.

[21] K.A. Lurie and A.V. Cherkaev, "The effective properties of composites and problems of optimal designs of constructions" (in Russian), Uspekhi Mekaniki (Advances in Mechanics) 2 (1987) 3-81.

[22] R.V. Kohn and G.W. Milton, "On bounding the effective conductivity of anisotropic composites," in *Homogenization and effective moduli of materials and media,* pp. 97-

125, ed. J.L. Ericksen, D. Kinderlehrer, R. Kohn and J.-L. Lions, Springer-Verlag, New York, 1986.

[23] G.W. Milton, "A brief review of the translation method for bounding effective elastic tensors of composites," to appear in the proceedings of the 6th Symposium on Continuum Models and Discrete Systems, ed. G.A. Maugin, Longman, 1990.

[24] G.A. Francfort and F. Murat, "Homogenization and optimal bounds in linear elasticity," Archives Rat. Mech. and Analysis 94 (1986) 307-334.

[25] Z. Hashin and S. Shtrikman, "A variational approach to the theory of the elastic behavior of multiphase materials," J. Mech. Phys. Solids 11, (1963) 127-140.

[26] G.W. Milton, "On characterizing the set of possible effective tensors of composites: the variational method and the translation method," Commun. Pure. Appl. Math 43 (1990) 63-125.

[27] G.W. Milton, "The coherent potential approximation is a realizable effective medium scheme," Commun. Math. Phys. 99 (1985) 463-500.

[28] M. Avellaneda, "Iterated homogenization, differential effective medium theory and applications," Commun. Pure Appl. Math. 40 (1987) 527.

[29] J. Berryman, "Effective medium theory for elastic composites," in *Elastic wave scattering and propagation*, p. 111, ed. V.K. Varadan and V.V. Varadan, Ann Arbor, MI, 1982.

[30] G.W. Milton, "Multicomponent composites, electrical networks and new types of continued fractions I and II," Commun. Math. Phys. 111 (1987), 281-327; 329-372.

[31] G. Allaire and R.V. Kohn, private communication, New York, 1990.

Graeme W. Milton
New York University
Courant Institute
251 Mercer Street
New York, NY 10012

Composite media and Dirichlet forms

UMBERTO MOSCO

Some relevant "macroscopic" features of bodies with complicated "microscopic" structure are usually described, in the mathematical theory of *composite media* and ***homogenization***, in terms of asymptotic properties of sequences of Dirichlet integrals

(1) $$E_h = \int_\Omega \sum_{ij=1}^{N} \partial_i u \; \partial_j u \; a_h^{ij}(x)dx \qquad , \quad h \in \mathbb{N},$$

$\partial_i = \partial u/\partial x_i$, $\partial_j = \partial u/\partial x_j$, by appropriately defining the "conductivity" coefficients $a_h^{ij}(x)$ on some open subset Ω of $\mathbb{R}^N$.

For every $h \in \mathbb{N}$, $(a_h^{ij}(x))$ is taken to be a non-negative definite, symmetric, N×N matrix defined on Ω . If, in addition, the sequence $(a_h^{ij}(x))$ is assumed to satisfy a uniform energy bound in Ω , and appropriate boundary conditions are prescribed on $\partial\Omega$, a sequence of selfadjoint operators in $L^2(\Omega)$ is associated with the functionals (1).

Then, a basic result of the theory affirms that an asymptotic "effective" conductivity matrix $(a^{ij}(x))$, and a corresponding " energy functional"

(2) $$E = \int_\Omega \sum_{ij=1}^{N} \partial_i u \; \partial_j u \; a^{ij}(x)dx \qquad ,$$

exist, satisfying the same bound in Ω, such that solutions, resolvent operators, eigenvalues and eigenfunctions of the boundary value problems associated with the sequence (1), as well as the corresponding evolution semigroups, have subsequences, as $h \to \infty$, that converge in appropriate sense to the solution, resolvent operator, etc., of the analogous boundary value problem in Ω for the selfadjoint operator associated with the asymptotic functional (2).

If the conductivities $a_h^{ij}(x)$ are allowed to develop some singularity in Ω as $h \to \infty$, the essential features of the theory, in particular the form (2) of the asymptotic energy, are not lost, provided a bound of the type

$$(3) \qquad 0 < \lambda|\xi|^2 \le \sum_{ij=1}^{N} a_h^{ij}(x)\, \xi_i \xi_j \le \Lambda b_h(x)|\xi|^2 \quad , x \in \Omega \ , \quad \xi \in \mathbb{R}^N ,$$

$0 < \lambda \le \Lambda$, keeps to hold for every h, with "weights" $b_h(x)$ satisfying

(4) $b_h(x)$ *converges weakly in* $L^1(\Omega)$ *as* $h \to \infty$.

However, if (3) holds with $b_h(x)$ just uniformly bounded in the $L^1(\Omega)$ norm, but not necessarily equi-integrable in Ω, then examples are known which show that the previous asymptotic description in terms of functionals of the form (2) *fails* in general. *Measure valued conductivities* $\nu^{ij}(dx)$ may occur in the limit, replacing the $a^{ij}(x)dx$ in (2), with singular components with respect to the N-dimensional Lebesgue measure, and even *non-local* terms may appear, destroying the property of functionals such as (2) of being additive with respect to the domain Ω.

One of these examples is the so called *highly conductive thin layer*. The conductivity

$$a_h^{ij}(x) = [1_{\Omega - \Sigma_h}(x) + ch\, 1_{\Sigma_h}(x)]\, \delta^{ij} \quad , \quad x \in \Omega \ ,$$

as $h \to \infty$ increases as ch on a layer Σ_h of decreasing width $1/h$ surrounding an (N–1)-dimensional surface Σ, while it is kept fixed equal to 1 in $\Omega-\Sigma_h$. The asymptotic energy takes the form

$$E = \int_\Omega [(\partial_1 u)^2+(\partial_2 u)^2+(\partial_3 u)^2]\, dx_1 dx_2 dx_3 + c \int_\Sigma [(\partial_1 u)^2+(\partial_2 u)^2]\, dx_1 dx_2 \quad ,$$

where for simplicity we have taken $N=3$ and Σ is the (x_1,x_2) plane. This can be re-written as

$$E = \int_\Omega \sum_{ij=1}^{N} \partial_i u\, \partial_j u\, \nu^{ij}(dx) \quad ,$$

with $\nu^{11} = \nu^{22} = dx_1 dx_2 dx_3 + c\, dx_1 dx_2\, \delta_{\{0\}}(dx_3)$, $\nu^{33} = dx_1 dx_2 dx_3$, $\nu^{ij} = 0$ if $i \neq j$.

We note, incidentally, that the surface energy term induces a *second order* transmission condition on the layer Σ, namely

$$[\partial u/\partial n_\Sigma] = c\, \Delta_\Sigma \qquad \text{on } \Sigma \ ,$$

where $[\partial u/\partial n_\Sigma]$ denotes the jump of the normal derivative of u accross Σ and Δ_Σ the "tangential" Laplacian on Σ. The corresponding diffusion process has been studied by Ikeda-Watanabe and, more recently, by Tomisaki.

A non-local example is obtained by taking conductivities

$$a_h^{ij}(x) = [|x-a|^{1-N} + |x-b|^{1-N} + ch^{N-1} 1_{T_h}(x)]\, \delta^{ij} \quad , \quad x \in \Omega \ ,$$

which are very singular at two distinct points a and b of Ω and along a thin tube T_h of thickness $1/h$ connecting a to b in Ω, *e.g.*, T_h is the neighbourhood of radius $1/h$ of the segment $[a,b] \subset \Omega$, $N \geq 4$. The asymptotic energy is now

$$E = \int_\Omega (|x-a|^{1-N} + |x-b|^{1-N})|Du|^2\, dx + \frac{c}{d}(u(a)-u(b))^2 \quad , \qquad d = b-a \ ,$$

and this can be also written in the form

$$E = \int_\Omega (|x-a|^{1-N} + |x-b|^{1-N})|Du|^2\, dx + \iint_{\Omega\times\Omega} (u(x)-u(y))^2\, j(dx,dy) \quad ,$$

where $j(dx,dy) = \frac{c}{d}\, \delta_{\{a,b\}}(dx,dy)$.

No general result seems to have been known so far that applies to wide classes of examples of this type. Some time ago John Baxter drew my attention to the beautiful theory of the ***Dirichlet forms*** of Beurling and Deny. Our aim here is to show that this theory provides a natural setting for the study of the asymptotic energies occurring in composite media.

We start by presenting a result that applies to arbitrary sequences of conductivity matrices $(a_h^{ij}(x))$ for which a bound like (3) holds, with weights $b_h(x)$ satisfying the condition

(5) $b_h(x)dx$ *converges weakly to a Radon measure* $\beta(dx)$ *in* Ω *as* $h\to\infty$.

Under the assumption (3) and (5), an asymptotic energy functional E always exists, of the form

$$(6)\; E = \int_\Omega \sum_{ij=1}^{N} \partial_i u \partial_j u\, \nu^{ij}(dx) + \int_\Omega u^2 k(dx) + \iint_{\Omega\times\Omega-\mathrm{diag}} (u(x)-u(y))^2\, j(dx,dy),$$

where $\nu^{ij}(dx)$ is a tensor-valued Radon measure in Ω that satisfies the condition

$$0 < \lambda|\xi|^2 \le \sum_{ij=1}^{N} \nu^{ij}(dx)\, \xi_i\xi_j \le \Lambda\beta(dx)|\xi|^2 \quad , \qquad \xi\in \mathbb{R}^N ,$$

$k(dx)$ is a positive Radon measure in Ω , $j(dx,dy)$ is a symmetric positive measure in $\Omega\times\Omega$ with support off the diagonal. With the functional (6) and the prescribed boundary condition on $\partial\Omega$, a positive self-adjoint operator A in $L^2(\Omega)$ is associated, whose domain contains $C_0^1(\Omega)$ and is contained in the

Sobolev space $H^1(\Omega)$. Solutions, resolvent operators, etc. of the boundary value problems for the sequence (1) have subsequences that converge as before in appropriate sense as $h \to \infty$ to the solution, resolvent operator, etc., of the boundary value problem in Ω for the operator A . The integral representation (6) extends to arbitrary u and v in the domain of the smallest self-adjoint extension of A with domain $C_0^1(\Omega)$, by taking quasi-continuous versions of the functions u(x), v(x) relative to the intrinsic capacity associated with E . Moreover, under the stronger assumptions (3) and (4), the representation (6) holds with $j \equiv 0, k \equiv 0$ and v^{ij} absolutely continuous with respect to dx.

An interesting special class of functionals (6), invariant under translations in $\mathbb{R}^N$, is the one of the form

$$E = \int_{\mathbb{R}^N} \sum_{ij=1}^{N} \partial_i u \partial_j u \, v^{ij} dx + \int_{\mathbb{R}^N} u^2 \, k dx + \iint_{\mathbb{R}^N \times \mathbb{R}^N - \mathrm{diag}} \frac{(u(x)-u(y))^2}{|x-y|^{N+\alpha}} \, c dx dy \ ,$$

with v^{ij} , k , c constants and $0 < \alpha < 2$, occuring in *homogenization models* in $\mathbb{R}^N$.

More general energy functionals of the type (6) can be considered, for instance, by allowing the conductivity coefficients v^{ij} to vanish, or to be infinite, on some subset E of Ω ; the measures k to be infinite on a subset F of Ω , as in [4] ; the measure j to be infinite on a subset G of $\Omega\times\Omega-$diag , as in [3], the sets E, F, G, being possibly fragmented and of complicated geometry.

This brings into the picture models of composite media with insulating inclusions E, or conductive inclusions E with constant potential on each connected component of E, inducing Neumann or transmission conditions on the boundary ; perfect conductive inclusions F forcing the potential to vanish on F and inducing a homogeneous Dirichlet condition on the boundary; symmetric "coupled" inclusions G forcing the potential to take the same value on each pair of matched components of G, with generalized "periodic" conditions on the boundary and a corresponding change in the topological type of the domain Ω.

We now outline the main features of an "abstract" unifying framework, based on the general theory of Dirichlet forms, referring to [5], [6] for more details. Such an abstract formulation appears to be in itself the natural one from the point of view of the variational convergence approach to composite media and homogenization. Moreover, it gives the analytic theory a form that reveals its underlying probabilistic structure and the intimate relationship with Markov processess, as in the special examples of [1]. It also provides intrinsic tools that can be used, for instance, when the state space Ω has the structure of a differentiable manifold, or it is just a discrete set.

We take X to be an arbitrary locally compact separable Hausdorff space and m a given positive Radon measure supported on the whole of X. By H we denote the Hilbert space $L^2(X,m)$ with inner product $(u,v) = \int_X uvm(dx)$.

A *form* E in H is intended to be a non-negative definite, symmetric bilinear form defined on a linear subspace $D[E]$ of H, the *domain* of E. A form E is said to be *closed* in H if $D[E]$ is complete under the inner product $E(u,v)+(u,v)$. A *Dirichlet form* in H is a densely defined, closed form E in H, such that $T \circ u \in D[E]$ and $E(T\circ u, T\circ u) \leq E(u,u)$, whenever $u \in D[E]$ and $T: \mathbb{R} \to \mathbb{R}$, $T(0) = 0$, $|T(x)-T(y)| \leq |x-y|$ for every $x, y \in \mathbb{R}$.

With every Dirichlet form E in H, a non-negative definite self-adjoint operator $-A$ is uniquely associated, the *generator* of E, such that $D[-A] = D[E]$, $E(u,v) = (\sqrt{-A}\, u, \sqrt{-A}\, v)$ for every $u, v \in D[E]$.

Dirichlet forms may initially be given, or represented, only on some linear subspace of their domain. In this respect the following notions are useful. A *Markovian form* in H is a form E in H, such that for every $\varepsilon > 0$ there exists $T_\varepsilon: \mathbb{R} \to [-\varepsilon, 1+\varepsilon]$, with $T_\varepsilon(t) \equiv t$ for $t \in [0,1]$ and $0 \leq T_\varepsilon(t') - T_\varepsilon(t) \leq t'-t$ for every $t'<t$, such that $T_\varepsilon \circ u \in D[E]$ and $E(T_\varepsilon \circ u, T_\varepsilon \circ u) \leq E(u,u)$ whenever $u \in D[E]$. A form E is *closable* in H if $(u_n) \subset D[E]$,

$E(u_n-u_m,u_n-u_m) \to 0$, $(u_n,u_n) \to 0$ as $n, m \to \infty$ imply $E(u_n,u_n) \to 0$ as $n \to \infty$. A form is closable in H if and only if there exists a closed extension of E in H, then there exists a smallest closed extension of E in H. A Dirichlet form is a Markovian form which is also densely defined and closed in H.

We say that a sequence of forms (E_h) *Γ-converges* to a form E in H if:

(a) for every $v_h \to u$ in H, $\liminf E_h(v_h,v_h) \geq E(u,u)$ as $h \to \infty$,

(b) for every $u \in H$, there exists $u_h \to u$ in H, such that

$$\limsup E_h(u_h,u_h) \leq E(u,u) \text{ as } h \to \infty,$$

where the quadratic forms involved are extended to be $+\infty$ on H outside their domain.

We also say that (E_h) *converges in the resolvent sense* to a form E in H if (a) holds for every v_h converging *weakly* to u in H and (b) holds as before.

A sequence (E_h) converges in the resolvent sense to E in H if and only if the *resolvent operators* $G_{h,\beta}$, $\beta>0$, associated with the forms E_h in H, converge to the resolvent operator G_β of the form E, as $h \to \infty$, in the strong operator topology.

We first state a compactness result for Markovian forms that satisfy an asymptotic "regularity" condition.

Let (E_h) be an arbitrary family of Markovian forms in H and let the following condition be satisfied:

There exists a dense subset $D \subset C_0(X)$, such that for every $u \in D$ we have

(7) $\liminf E_h(u_h,u_h) < +\infty$ as $h \to \infty$ for some $u_h \to u$ in H as $h \to \infty$.

Then a subsequence $(E_{h'})$ exists that Γ-converges, as $h' \to \infty$, to a Dirichlet form E in H, whose domain contains D, and we have

$$(8) \qquad E = \int_X \mu_{<u,v>}(dx) + \int_X \tilde{u}\tilde{v}\, k(dx) + \iint_{X\times X-\mathrm{diag}} (\tilde{u}(x)-\tilde{u}(y))(\tilde{v}(x)-\tilde{v}(y))\, j(dx,dy)$$

for every u,v belonging to the domain of the smallest closed extension of $E|_D$, the restriction of E to D. In (8), j(dx,dy) is a symmetric non-negative Radon measure on X×X off the diagonal, which does not charge any subset of X×X–diag whose projection on the factor X has E-capacity zero, k(dx) is a non-negative Radon measure in X not charging any set of E-capacity zero, $\mu_{<u,v>}(dx)$ is a Radon measure-valued non-negative definite symmetric bilinear form in u, v. If, in addition, X has the structure of a differentiable manifold, with local coordinates belonging to the domain of the form, then we have

$$\mu_{<u,v>} = \sum_{ij=1}^{N} \partial_i u \partial_j v\, v^{ij} , \qquad v^{ij} = \mu_{<x_i,x_j>} ,$$

and v^{ij} is an invariantly defined symmetric tensor-valued measure on X, satisfying

$$0 \le \sum_{ij=1}^{N} v^{ij} \xi_i \xi_j \qquad \text{for every } \xi \in \mathbb{R}^N .$$

This result can be applied to sequences of Markovian forms of the type

$$(9) \qquad E_h = \int_X \mu_{h,<u,v>}(dx) + \int_X \tilde{u}\tilde{v}\, k_h(dx) + \iint_{X\times X-\mathrm{diag}} (\tilde{u}(x)-\tilde{u}(y))\,(\tilde{v}(x)-\tilde{v}(y))\, j_h(dx,dy),$$

where for every h ,

$$E_h^{(c)} = \int_X \mu_{h,<u,v>}(dx)$$

is a densely defined closable Markovian form in H, k_h is a positive Borel measure on X not charging subsets of X of $E_h^{(c)}$-capacity zero, but possibly $+\infty$ on large subsets of X, j_h is a symmetric positive Borel measure on X×X–diag, possibly $+\infty$ valued, not charging subsets of X×X–diag whose projection on X has $E_h^{(c)}$-capacity zero and where $\tilde{u}$ and $\tilde{v}$ are $E_h^{(c)}$-quasi continuous versions of u, $v \in D[E_h^{(c)}]$. Forms of this type may be useful to describe wide classes of composite media with inclusions, as we have already mentioned.

Related to the previous result is the following compactness theorem for Dirichlet forms of diffusion type. Let

$$E_h = \int_X \mu_{h,<u,v>}(dx) \quad , \quad h \in \mathbb{N} ,$$

be a sequence of closable Markovian forms with domain $D[E_h] = D$, where D is a subalgebra of $C_0(X)$ that is separating in the following sense: For every compact subset V of X contained in a relatively compact open subset U of X, there exists a non-negative function $\alpha \in D$, such that $\alpha(x) = 1$ if $x \in V$ and $\alpha(x) = 0$ if $x \in X-U$. For every $u \in D$ let the measures $\mu_{h,<u,v>}$ be absolutely continuous with respect to m , uniformly in h , and let the sequence $(\mu_{h,<u,u>})$ be bounded. Then, a subsequence $(E_{h'})$ exists that Γ-converges as $h' \to \infty$ to a Dirichlet form E in H whose domain contains D , and E is of diffusion type

$$E^{(c)} = \int_X \mu_{<u,v>}(dx) , \tag{10}$$

with the measure $\mu_{<u,v>}$ absolutely continuous with respect to m , for every (bounded) u,v in the domain of the smallest closed extension of $E|_D$.

When the asymptotic regularity assumption (7) is not satisfied, the representation of the limit form presents additional technical difficulties.

The following result can be considered as a general compactness result for "killing" measures k_h as those occurring in (9). Let (E_h) be a sequence of Markovian forms in H, of the type

$$E_h = \int_X \mu_{h,<u,v>}(dx) + \int_X \tilde{u}\tilde{v}\, k_h(dx) \quad ,$$

where for every h

$$E_h^{(c)} = \int_X \mu_{h,<u,v>}(dx)$$

is a closable Markovian form in H with domain $D[E_h^{(c)}] = D$, D dense in $C_0(X)$, and k_h is an arbitrary positive, possibly $+\infty$ valued, Borel measure on X, not charging sets of $E_h^{(c)}$-capacity zero. Let the sequence $(E_h^{(c)})$ Γ-converge to a Dirichlet form $E^{(c)}$ in H, of the diffusion type (10), with $D[E^{(c)}]\supset D$. Then, a subsequence $(E_{h'})$ exists that Γ-converges as $h' \to \infty$ to the form

$$E = \int_X \mu_{<u,v>}(dx) + \int_X \tilde{u}\tilde{v}\, k(dx) \;,$$

where k is some real extended positive Borel measure on X, null on subsets of $E^{(c)}$-capacity zero.

Similar compactness results hold for families of "jumping" measures j_h as those occurring in (9), provided the support of j_h stays off the diagonal of $X \times X$ as $h \to \infty$.

All the results stated above can be easily converted from Γ-convergence to convergence in the resolvent sense, provided we assume in addition that $D[E_h]\subset K$ for every h, with equicontinuous injections, where K is a *compact* subspace of H. More generally, it suffices to assume that for every c>0 the level sets $\{u \in D[E_h] : E_h(u,u) \le c\}$ are continuously injected in K, uniformly in h.

The crucial property in the preceding theory is the closability of the forms involved. Let us just point out that this notion is related to the possibility of defining *traces* and suitable *generalized gradients* of the functions in the domain of the *closed* extensions of the forms. For an interesting class of forms, as for instance those related to the so called highly insulating thin layers, these generalized gradients can be defined as measures, introducing suitable BV spaces , as the Special BV spaces recently studied by De Giorgi, Ambrosio et al.

Finally, we remark that the study of the *regularity* and local behaviour of minimizers of functionals such as (6) is largely open and presents substantial difficulties, since *measure coefficients* will in general take the place of the bounded measurable coefficients of the De Giorgi-Nash theory. Intrinsic metrics and capacities should be expected to be useful tools. For a special class of "weighted" degenerate elliptic equations, related to the non-local examples mentioned before, intrinsic capacities have been introduced by Fabes-Jerison-Kenig and applied in [2] to get pointwise estimates for the decay of the solutions.

Acknowledgments. I wish to thank the Institut für Angewandte Mathematik SFB 256 of the University of Bonn, where this research was carried out, for hospitality and support.

REFERENCES

[1] J.R. Baxter, G. Dal Maso, U. Mosco, *Stopping times and Γ-convergence*, Trans. AMS 303 (1987), 1-38.

[2] M. Biroli, U. Mosco, *Wiener criterion and potential estimates for obstacle problems relative to degenerate elliptic operators*, Universität

Bonn SFB 256 Preprint Series N.69 (1989), to appear in Ann. Mat. Pura Appl. (4).

[3] R. Gulliver, G. Dal Maso, U. Mosco, *Asymptotic spectrum of manifolds of increasing topological type*, Universität Bonn SFB 256 Preprint Series, to appear.

[4] G. Dal Maso, U. Mosco, *Wiener's criterion and Γ-convergence*, J. Appl. Math. and Opt. **15** (1987), 15-63.

[5] U. Mosco, *Composite media and asymptotic Dirichlet forms*, to appear.

[6] U. Mosco, *Compact families of Dirichlet forms*, to appear.

Umberto Mosco
Dipartimento di Matematica
Università "La Sapienza"
I-00185 Roma

Progress in Nonlinear Differential Equations and Their Applications

Editor
Haim Brezis
Département de Mathématiques
Université P. et M. Curie
4, Place Jussieu
75252 Paris Cedex 05
France
and
Department of Mathematics
Rutgers University
New Brunswick, NJ 08903
U.S.A.

Progress in Nonlinear Differential Equations and Their Applications is a book series that lies at the interface of pure and applied mathematics. Many differential equations are motivated by problems arising in diversified fields such as Mechanics, Physics, Differential Geometry, Engineering, Control Theory, Biology, and Economics. This series is open to both the theoretical and applied aspects, hopefully stimulating a fruitful interaction between the two sides. It will publish monographs, polished notes arising from lectures and seminars, graduate level texts, and proceedings of focused and refereed conferences.

We encourage preparation of manuscripts in some form of TEX for delivery in camera ready copy, which leads to rapid publication, or in electronic form for interfacing with laser printers or typesetters.

Proposals should be sent directly to the editor or to: Birkhäuser Boston, 675 Massachusetts Avenue, Suite 601, Cambridge, MA 02139.

PNLDE1 Partial Differential Equations and the Calculus of Variations, Volume I
Essays in Honor of Ennio De Giorgi
F. Colombini, A. Marino, L. Modica, and S. Spagnolo, editors

PNLDE2 Partial Differential Equations and the Calculus of Variations, Volume II
Essays in Honor of Ennio De Giorgi
F. Colombini, A. Marino, L. Modica, and S. Spagnolo, editors

PNLDE3 Propagation and Interaction of Singularities in Nonlinear Hyperbolic Problems
Michael Beals

PNLDE4 Variational Methods
Henri Berestycki, Jean-Michel Coron, and Ivar Ekeland, editors

PNLDE5 Composite Media and Homogenization Theory
Gianni Dal Maso and Gian Fausto Dell'Antonio, editors